PUBLICATIONS OF THE ISRAEL ACADEMY

OF SCIENCES AND HUMANITIES

SECTION OF SCIENCES

———

FAUNA PALAESTINA

INSECTA V — ODONATA OF THE LEVANT

FAUNA PALAESTINA

Pseudagrion sublacteum mortoni (foreground), a Levantine endemic, and
Trithemis arteriosa, an afrotropical species widespread in semi-arid environments

FAUNA PALAESTINA · INSECTA V

ODONATA OF THE LEVANT

by

HENRI J. DUMONT

Jerusalem 1991

The Israel Academy of Sciences and Humanities

Author's Address:
Institute of Ecology
University of Ghent
K. Ledeganckstraat 35
B-9000 Ghent, Belgium

ISBN 965-208-013-6
ISBN 965-208-097-7

Printed in Israel
at Keterpress Enterprises, Jerusalem

CONTENTS

PREFACE

The Odonata fauna of the Levant dealt with in this volume covers 82 species, belonging to 36 genera of the suborders Zygoptera (damselflies) and Anisoptera (dragonflies).

The geographical scope is not restricted to cis- and trans-Jordan, and information is given on species found in Israel, Egypt (Sinai), Jordan, Syria and the Lebanon.

The main repositories are the National Entomological Collection at Tel Aviv University, the "Beth Gordon" Collection at Kibbutz Deganya A, my personal collections, Armin Heymer's collection from Israel and Sinai, the British Museum (Natural History), and the Royal Scottish Museum, Edinburgh.

Larval forms are discussed in a special section, including identification keys.

The numbers in parentheses which appear after the locations refer to the geographical areas on the map of Israel and Sinai, at the end of the book. The spelling of names of localities in Israel and Sinai is according to the maps published by the Survey of Israel.

1

ACKNOWLEDGEMENTS

I dedicate this book to my children Patrick and Jani who, for many years, knew their father as that strange visitor to the house, who used to rush in to drop one suitcase, pick up another, and again depart for some mysterious destination, a desert or some other locality. I hope it may show them that this was not all just for fun or because of some weird, and as I now experience, highly contagious travel virus.

The field work involved in this study was supported by the Israel Academy of Sciences and Humanities, but part of the data were also collected in the framework of grant 2.0009/75, awarded to me by the Fonds voor Kollektief Fundamenteel Onderzoek (National Science Foundation), Belgium. In Israel, I was generously received by Prof. F.D. Por (Jerusalem) and Prof. J. Kugler (Tel Aviv), and shown around most ably by Mr M. Samocha.

During an additional field trip in Jordan, I enjoyed the company and assistance of my old travel companion, Mr Jo Vermeir.

The text was typed by Nicole Declercq, but the finishing touch was, of course, given by Simonne Wellekens.

I am grateful to Patrick Dumont for the lovely colour plate he painted for this book, and to Jeannine Pensaert and Marcel Bruyneel for their skilful input into the final form of the illustrations.

Philip Corbet, Charles Degrange, Heinrich Lohman, Hans Klaus Pfau and Wolfgang Schneider kindly granted permission to reproduce some of their figures, as did the Societas Internationalis Odonatologica and its editor, Bostjan Kiauta. Armin Heymer donated rich dragonfly collections made by him in Israel and Sinai, and Cynthia Longfield kindly sent lists of Middle Eastern Odonata in the Collections of the British Museum (Natural History). While preparing this text, I benefitted from exchanging opinions with such grey eminences of Odonatology as Maurits Lieftinck and Elliot Pinhey, but I also enjoyed numerous discussions with young, talented dragonfly workers such as Wolfgang Schneider and Koen Martens.

Koen Maes assisted with the microphotographical work, and Simonne Wellekens and the technical staff of ISI Benelux helped with the scanning electron microscope work, most of which was done using a type SX 30 of ISI.

I particularly thank F.D. Por and for that matter, the entire *Fauna Palaestina* Committee, for their endless patience in waiting for the present text. It took several years to print the book, a frustrating exercise in patience for an author. The result, which is the outcome of a meticulous editing job by Ms Ilana Ferber and Mr R. Amoils, undoubtedly improved the quality of the work, and was therefore well worth waiting for.

Ms Judith Kahan and Ms Tamara Leff did the typesetting and Ms Ferber also prepared the index and saw the book through the press.

LIST OF ABBREVIATIONS

A – anal vein or analis;
Ab – anal bridge;
A.C. – anteclypeus;
Ac – anal crossing;
ac – anterior collar;
AFr (=VR) – anterior frame;
ah. str. – antehumeral stripe;
al. sin. – alar sinus;
AK – ejaculation chamber;
an – antenodal cross-veins;
Anus – anal opening;
A.O. – anterior ocellus;
$Ap_{1,2}$ – apodemes;
app. dors – appendix dorsalis or lamina supra analis;
arc – arculus;
At – anal triangle;
aw – anterior wing implant;
b – bursa;
ba – bursal arms;
BHA – base of anterior hamulus;
BV – base of vesica spermalis;
C – costal vein or costa;
car – carina;
cerci – appendices superiores;
cf – carinal fork;
Cu – cubital vein or cubitus;
Cu_1 and Cu_2 – first and second branches of cubital vein;
Cv – cubital cross-veins;
Cx_1 – coxa of first pair of legs;
Cx_2 – coxa of leg 2;
Cx_3 – coxa of leg 3;
d – discoidal cell (quadrangle);
d – discoidal cell (wing triangle);
epm_2 – mesoepimerum;
epm_3 – metaepimerum;
$epst_2$ – mesoepisternum;
$epst_3$ – metaepisternum;

F – frons;
f.g. – female gonapophyses;
G – gena, genae;
gen.op. – genital opening;
Gl – glossa;
h – hook;
HA – hamulus anterior;
HP – hamulus posterior;
h.s. – humeral suture;
h. str. – humeral stripe;
ht – hypertriangle;
inf_2 – mesoinfraepisternum;
inf_3 – metainfraepisternum;
IP – inner plate of hamulus anterior;
IR_2 – second interradial vein;
IR_3 – third interradial vein;
L – ligula;
$L_{1,2}$ – segments of ligula;
LA – lamina anterior;
LB – lamina batilliformis;
lam. mes. – lamina mesostigmalis;
lam. sub. an. – lamina subanalis (= appendix inferior);
lam. supra an. – lamina supra-analis;
Lbr – labrum;
lr – lower rim of hind margin of pronotum;
m – caudal membranaceous part of vesica spermalis;
Md – mandibula, mandibles;
me_1, me_2 – median lobes of pronotum;
M – median vein or Medius;
Me – membranula;
MR – opening of reservoir of vesica spermalis;
ms – mesostigma [= s2];
Mspl – median supplemetary sector (vein);
mt – metasternum;
Mx – maxilla
N – nodus;

Occ – occiput;

os – ostium (opening of bursa);

P (=Z) – plug;

P_1 – filling pore [vesica spermalis];

P_1 – first segment of paraglossa;

P_2 – second segment of paraglossa or palpus;

P.C. – postclypeus;

Pe – ejaculatory pore;

PF – processus furculiformis;

PFr (=HR) – posterior supporting frame of hamuli;

Pn – pronotum;

pn – postnodal cross-veins;

P.Oc. – postocular spots;

P.Oc.T. – postocellar triangle;

PsPt – pseudopterostigma;

pr – posterior rim of pronotum;

Pt – pterostigma;

pw – posterior wing implant;

r – sperm reservoir;

R – radial vein or Radius;

R_1 – R_{4+5} – branches of radius;

R + M – fused radius and medius, basal to arculus;

Rf – radial fork;

Rs – radial sector;

Rspl – radial supplementary sector;

s_2 – mesostigma;

s_3 – metastigma;

Sarc – sector of the arculus;

Sc – subcostal vein;

$Schw_v$ – erectile tissue;

St – styli;

stm – stigma;

str – sterigmata (sclerotized plates);

sts – stylets;

Su_1 – suture 1;

Su_2 – suture 2;

ur – upper rim of hind margin of pronotum;

V – vertex;

V_{1-4} – segments of vesica spermalis;

V_{1-3} – first to third pair of valves [ovipositor];

V_a – apical valves;

V_b – basal valves;

Vf – valvifer;

vl – vulvar lips;

VS – vesica spermalis;

V.v – vulvar scales or lips (valvulae vulvae).

4

INTRODUCTION

GENERAL CHARACTERS OF THE ORDER ODONATA

Amphibiotic, hemimetabolous, medium- to large-sized four-winged insects. Imaginal state comparatively short (from a few days to two months), aerial; larval stage usually much longer-lived, aquatic. All stages are predacious. Compound eyes large. Antennae short. Prothorax small, free.

Meso- and metathorax fused, tilted at an angle to the body axis, so that the legs are moved frontad and the wings posteriad. Abdomen composed of 11 segments. Male with reduced penis, but with a secondary copulatory organ and accessory genitalia on ventrum of abdominal segments 2 and 3. Females oviparous, with or without an ovipositor.

Sexes of similar size and habitus, but often differing in colour pattern of the thorax, abdomen, legs and/or wings.

Three suborders are extant. Of these, the Anisozygoptera are represented by one genus (*Epiophlebia*) and two species (one in Japan, one in the Central Himalayas).

Differences between suborders Zygoptera (the most primitive of all, and probably a monophyletic group) and Anisoptera (more advanced Odonata, but probably not a monophyletic group) will be discussed under the various sections on anatomy.

EXTERNAL ANATOMY OF THE ADULT

Head (Figs. 1–8): Large, extremely mobile, wider than thorax. Shape different in the two suborders (Zygoptera — transversely elongate; Anisoptera — subglobular), but also major differences among families. In the Zygoptera, the compound eyes are always widely separated from one another. Three ocelli (supplementary, single ommatidia) are situated on the vertex, here a flat area between the compound eyes. They are arranged in a triangle, and the foremost ocellus is slightly larger than the posterior pair. The antennae are short and consist of 7 segments. Scapus and pedicellum are usually more strongly built than the apical flagellum. The occiput is narrow medially, but expands laterally. When the dorsum of the head is dark coloured, one frequently observes two clear spots on either side of the occiput — the postocular spots. Anteriorly situated are the vertex, the frons which is rarely (*Ceriagrion*) transversely ridged, and the clypeus (epistome). The clypeus is divided into a large postclypeus, and a smaller, often triangular anteclypeus, adjacent to the labrum. The labrum overhangs the mouth. The mouth parts are raptorial. The mandibles are strongly developed; the number and shape of their teeth is genus-specific, often

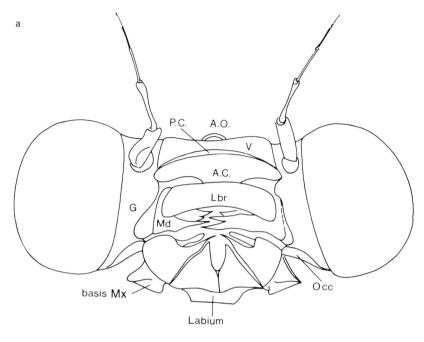

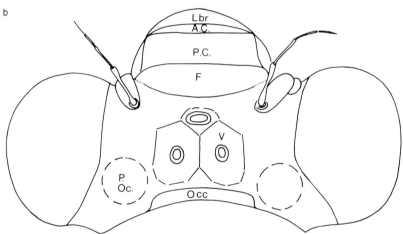

Fig. 1: Head of a zygopteran; a. ventral view; b. dorsal view
A.C. – anteclypeus; A.O. – anterior ocellus; basis Mx – base of maxilla;
F – frons; G – gena; Lbr – labrum; Md – mandibles; Occ – occiput;
P.C. – postclypeus; P.Oc. – postocular spots; V – vertex

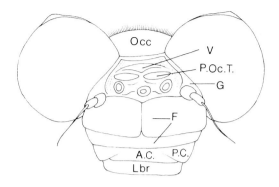

Fig. 2: Head of a gomphid
A.C. – anteclypeus; F – frons; G – genae; Lbr – labrum; Occ – occiput;
P.C. – postclypeus; P.Oc.T. – postocellar triangle; V – vertex

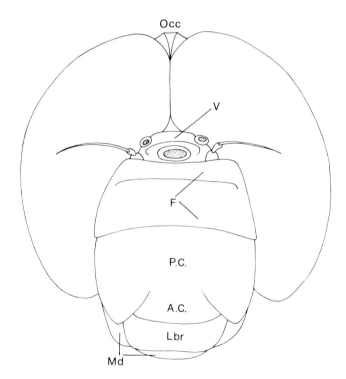

Fig. 3: Head of an aeschnid
A.C. – anteclypeus; F – frons; Lbr – labrum; Md – mandibula;
Occ – occiput; P.C. – postclypeus; V – vertex

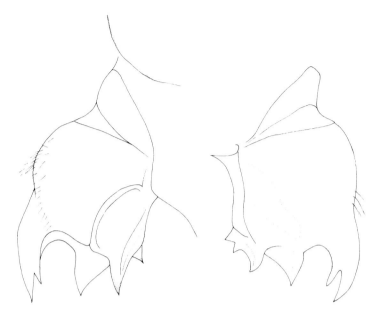

Fig. 4: Mandibles

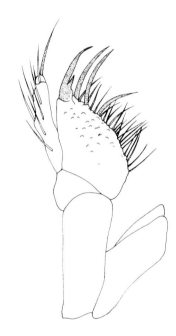

Fig. 5: Maxilla

species-specific (Fig. 4). Maxilla (Fig. 5) with two apical lobes, of which the internal one is strongest and bears rows of stiff internal spines. The external lobe is sometimes considered a palpus. The base is formed of the cardo and stipes, as usual.

Labium (Figs. 6–8) strongly developed. The glossae (internal lobi) are free in the Zygoptera (Fig. 6), or fused to form a single ligula in the Anisoptera (Fig. 7). A rudiment of a median cleft may, however, be present. It is relatively deep in the Cordulegasteridae (Fig. 8). The paraglossae (lateral lobi) are more strongly developed in the Anisoptera, and particularly in the Libelluloidea (= Libellulidae plus Corduliidae), than in the Zygoptera. They are composed of one or two articles. Again, the apical article is sometimes considered a palpus.

The anisopteran head is further distinguished by the fact that the eyes are larger than in the Zygoptera, often confluent along the mid-dorsum of the head (Libellulidae, Aeschnidae). In the Cordulegasteridae, the eyes meet only in one point, and in the Gomphidae they are separated (Fig. 2). However, in all Anisoptera the occiput is small, often reduced to a small triangle, and never much wider than the distance between the outer ocelli. The vertex, on which the ocelli are located, is usually differentiated, rarely flat. In forms where the compound eyes meet, the ocelli are arranged in a triangle as in Zygoptera. In most Gomphidae, however, they are more or less aligned.

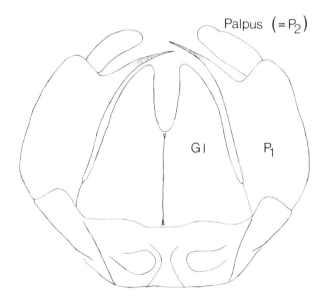

Fig. 6: Labrum of a zygopteran (*Calopteryx*)
Gl – glossa; P_1 – first segment of paraglossa;
P_2 – second segment of paraglossa or palpus

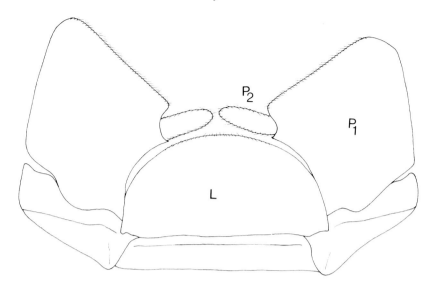

Fig. 7: Labrum of a (non-cordulegaster) anisopteran
L – ligula; P_1 – paraglossa; P_2 – paraglossa 2 or palpus

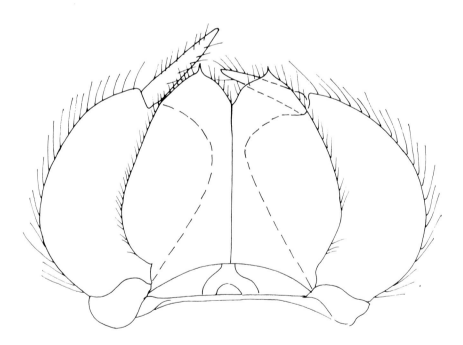

Fig. 8: Labrum of *Cordulegaster*, with median cleft in ligula

Thorax (Figs. 9–12): The prothorax is small and free. Its dorsum, the pronotum, is composed of an anterior collar, two median lobes, and a posterior rim. The posterior rim may be raised, and present various differentiations. Often, a median outgrowth is found, which may or may not be flanked by upright teeth, and flanked anteriorly by projecting stylets. In Zygoptera — particularly females — this, in conjunction with the lamina mesostigmalis, forms the grasping area for the male's anal appendages, and has enormous diagnostic value (see keys).

The meso- and metathorax are strongly developed and fused into a single block — the synthorax or pterothorax. This "block" stands obliquely, in an aerodynamic position, to the longitudinal body axis, so that the legs are moved frontad and the wings caudad. The "body" of the synthorax is formed by pleurites, including the anterior half of its dorsum. Tergites are small, reduced to a meso- and metascutum and scutellum, flanked by the membranaceous implants of the wings. The meso- and metasternum may present distinctive colour patterns.

Each of the strongly expanded pleurites is divided into three sclerites: an episternum, an epimerum, and an infra-episternum. The mesoepisterna ($epst_2$) on either side fuse along the mid-dorsum to form the median suture, often raised into a crest. Left and right rami of the crest usually remain visible as individual thickenings. Anteriorly, the median crest (carina) opens to form a median frame or fork, while posteriorly it forms two ante-alar sinuses. The carina may present a hump or a spine in Anisoptera (Fig. 10). The suture between $epst_2$ and epm_2 (= epimerum 2 or mesoepimerum) is called the humeral suture; it either runs straight or is somewhat wavy, but is always well developed and entire.

Clear bands on $epst_2$, flanked by black stripes on the carina and the humeral suture, are called antehumeral stripes. The dark striae overlying the humeral suture itself are the humeral bands. The humeral suture forks anteriorly to accommodate the infra-episternum 2 or mesoinfra-episternum (inf_2). The $epst_2$ has an anterior differentiation flanking the carinal frame — the lamina mesostigmalis (lam. mes.) (Fig. 11). In zygopteran copulation, it forms a functional unity with the hind lobe of the pronotum. It is the main area where the male superior appendages grasp. In some cases, areas of the largely membranaceous pre-episternum (that articulates with the prothorax and accommodates the mesothoracic stigmata) may be sclerotized (epaulettes) and take part in tandem linkage mechanics.

The suture (Su_1) between meso- and metathorax is complete in the Calopterygidae only (Fig. 9). In other Zygoptera and in all Anisoptera, it is only very partially visible. The suture (Su_2) between $epst_3$ and epm_3 is entire, anteriorly forked at the level of infra-episternum 3 (inf_3). Behind epm_3, an upfolded rim of the metasternum (mt) may be laterally visible.

The legs (Fig. 12) are prehensile. The prothoracic pair is shortest, the metathoracic pair longest. They are of classical structure: coxa, trochanter, femur, tibia, tarsus (3 segments) and claws. Both the shape and colour of parts of the legs (usually the tibiae) may be of taxonomical significance. Also the number, shape, and structure of the spines on the femur and tibia can vary with taxa at different levels.

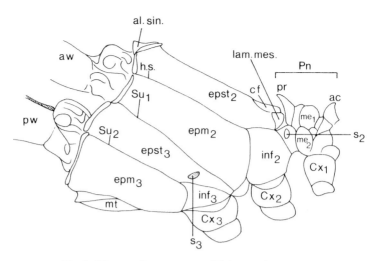

Fig. 9: Thorax of a zygopteran (*Calopteryx*)
ac – anterior collar of prothorax; al. sin. – alar sinus;
aw – anterior wing implant; cf – carinal fork;
Cx_1 – coxa of first pair of legs; Cx_2 – coxa of leg 2; Cx_3 – coxa of leg 3;
epm_2 – mesoepimerum; epm_3 – metaepimerum; $epst_2$ – mesoepisternum;
$epst_3$ – metaepisternum; h.s. – humeral suture; inf_2 – mesoinfraepisternum;
inf_3 – metainfraepisternum; lam. mes. – lamina mesostigmalis;
me_1, me_2 – median lobes of pronotum; mt – metasternum; Pn – pronotum;
pr – posterior rim of pronotum; pw – posterior wing implant;
s_2 – mesostigma; s_3 – metastigma; Su_1 – suture 1; Su_2 – suture 2

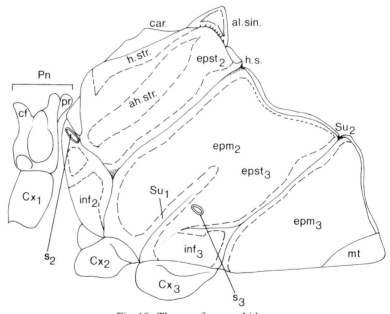

Fig. 10: Thorax of a gomphid
ah. str. – antehumeral stripe; car – carina; h. str. – humeral stripe;
All other symbols are as for Fig. 9

12

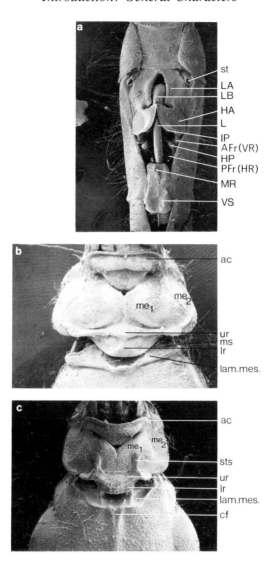

Fig. 11: a. Accessory genitalia of a coenagrionid
AFr (=VR) – anterior supporting frame of hamuli; HA – hamulus anterior;
HP – hamulus posterior; IP – inner plate of hamulus anterior; L – ligula;
LA – lamina anterior; LB – lamina batilliformis; MR – filling hole of
vesica spermalis; PFr (=HR) – posterior supporting frame of hamuli;
st – stigma; VS – vesica spermalis;
b–c. Pronotum structure in Zygoptera
ac – anterior collar; cf – carinal fork; lam. mes – lamina mesostigmalis;
lr – lower rim of hind margin of pronotum; me_{1-2} – median lobes;
ms – mesostigma; sts – stylets; ur – upper rim of hind margin of pronotum

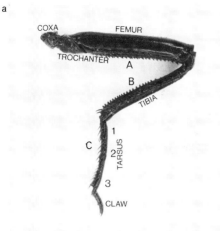

Fig. 12: a. Leg of a *Cordulegaster*
Note different types of ornamentation
(A) short, stiff spines; (B) knob-shaped sclerotizations; (C) long spines
b–c. Legs of *Platycnemis*
b. with tibia flattened but not expanded (*P. kervillei*);
c. with tibia flattened and expanded (*P. dealbata*)

Wings (Figs. 13–15): The wing is a double membrane, supported by a skeleton of 6 longitudinal, strongly sclerotized main veins, with numerous ramifications and cross-veins. The longitudinal veins all lie in different planes, so that the fractions of the membrane stretched between them stand at angles to one another. Wings may be hyaline, tinged, or display well-defined colour-spots. Wings may become darker with age, and become suffused with colour, especially around the veins. In perching, Zygoptera hold their wings closely apposed across the dorsum of the abdomen (or partially open in the genus *Lestes*). Anisoptera never close their wings when at rest, but the angle between the left and right pairs may vary according to families and genera.

The wing venation of dragonflies is complex but easy to study and has thus traditionally been used as a basis for classification and for discussing the phylogeny

14

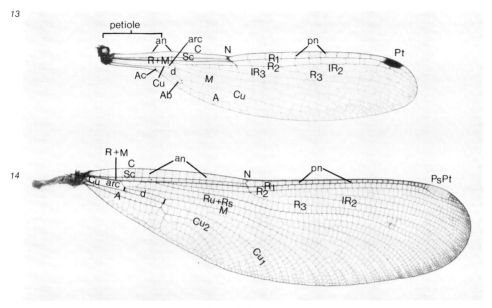

Figs. 13–14: Wings of Zygoptera
13. *Coenagrion*; 14. *Calopteryx*

A – anal vein or analis; Ab – anal bridge; Ac – anal crossing;
an – antenodal cross-veins; arc – arculus; C – costal vein or costa;
Cu – cubital vein or cubitus; Cu_1 and Cu_2 – first and second branches of
cubital vein; d – discoidal cell (quadrangle); IR_2 – second interradial vein;
IR_3 – third interradial vein; M – median vein or Medius; N – nodus;
pn – postnodal cross-veins; Pt – pterostigma; PsPt – pseudopterostigma;
R – radial vein or Radius; R_1 – R_{4+5} – branches of radius;
R + M – fused radius and medius, basal to arculus; Sc – subcostal vein

of the order. In the past fifteen years, however, alternative approaches have been made (see further). A nomenclature of the wing venation was devised by Comstock & Needham (1903). It was followed in early handbooks by Ris (1909–1919), Schmidt (1929) and May (1933). Tillyard (1917) and Tillyard & Fraser (1938–1940) later proposed an alternative system, which is now widely followed, and employed in this book.

The costa (C, or costal vein) forms the anterior border of the wing, from base to apex. The subcosta (Sc) extends from the base to a thick cross-vein (the nodus, N), somewhat before halfway along the length of the wing. The Radius (R) extends from wing base to apex (posterior to the nodus it is called R_1). Posterior to the nodus, there is one row of cells between C and R_1. Subapically, a large cell (extending over the length of a single or several normal cells) is found — the pterostigma (Pt). The Pt is often conspicuously coloured, and without cross-veins. Cross-veins are found only in the females of *Calopteryx*, where the structure is called the pseudopterostigma. *Calopteryx* males have no pterostigma at all.

15

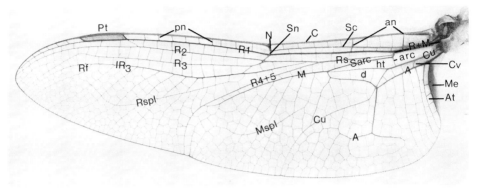

Fig. 15: Hind wing of anisopteran (*Anax*)
arc – arculus; At – anal triangle; Cv – cubital cross-veins;
d – discoidal cell (wing triangle); ht – hypertriangle; Me – membranula;
Mspl – median supplemetary sector (vein); Rf – radial fork; Rs – radial sector;
Rspl – radial supplementary sector; Sarc – sector of the arculus; sn – subnodus
Other symbols as for Figs. 13–14

The Medius (M) and the Radius are fused (R + M) between the base of the wing and a cross-vein, the arculus (arc), which is the site where M (and Rs — see hereafter) abruptly turn down. Via an extension of arc, the arculus sector (Sarc), M then continues its way towards the lower rim of the wing, which it reaches at about its middle. Next comes the cubitus (Cu), straight between the wing base and arc and then, either in a curve or in an angle, making up part of d (see hereafter), towards the wing rim. The last longitudinal vein is the analis (A). In Zygoptera with stalked wings (thus not in Calopterygidae and Euphaeidae), it builds the posterior rim of the petiole. In the two families cited, plus all Anisoptera, it may give rise to one or more branches, which may form loops and thus define particular cell groups (anal cells). In anisopteran hind wings, which are always broader than the forewings, a small group of cells, the anal triangle (At), is isolated between the proximal branch of A and the proximal wing margin. Adjacent to it lies a basal, usually coloured membrane, the membranula (Me). A membranula is never found in Zygoptera, where forewings and hind wings are always similar in shape.

On the arculus, above Sarc, springs another vein or sector, the radial sector (Rs). From this vein emanate all remaining branches of the radius: R_2, R_3, and R_{4+5} which remain fused in all modern Odonata. Between R_3 and R_{4+5} lies IR_3, the third interradial vein. In some of the larger Aeschnidae, IR_3 may be forked (Rf) near the level of the pterostigma. Not directly connected to any of the main veins are Rspl, the radial supplementary sector, and Mspl, the median supplementary sector. Fairly close to the wing base is located the discoidal cell (d). In Anisoptera, it is divided into the discoidal cell proper (d), and the hypertrigone or hypertriangle (ht). In Zygoptera, no hypertrigone is found and d consists of a single cell, except in the Calopterygidae, where d is elongate and traversed by a number of cross-veins.

16

Certain cross-veins are of particular importance: the antenodal cross-veins (an) extend from C and Sc to R (or R + M) between the wing base and the nodus (and its extension, the subnodus, Sn). In primitive Zygoptera, they are two in number, but more numerous in Calopterygidae and Euphaeidae. In Anisoptera, there are always more than two antenodals. However, in some less advanced families, the original two antenodals (called primary antenodals) can still be identified as more strongly sclerotized than the secondary antenodals. In evolved groups, like the Libellulidae, the primaries are completely merged with the secondary antenodals. By analogy, the cross-veins between C and R_1 are called postnodal cross-veins (pn).

Between Cu and A, from wing base to d, one or more cubital cross-veins (Cv) are found. In the more primitive Zygoptera, where Cv is usually a single vein, it is called anal crossing (Ac) and represents an important vestigial structure, indicating a former ending of the anal vein. Its point of insertion relative to the point where the anal vein branches off the wing petiole is of diagnostic value. The portion of the anal vein between the wing margin and a cross-vein situated at the distal corner of d is called the anal bridge (Ab) in Zygoptera.

Abdomen (Figs. 16–29): The abdomen consists of 10 complete segments, and a vestigial 11th one. It is elongated, cylindrical (all Zygoptera, non-libelluloid Anisoptera) or triangular (most Libellulidae) in cross-section, and in the latter case, often has a dorsal and two lateral carinae. The tergites are strongly developed, and form the dorsum and the sides of the segments; the pleurites are small and infolded; the sternites form the floor of the segments. The first segment is short, the second one longer, segments 3–7 (S_{3-7}) long, S_8 and S_9 again shorter, and S_{10} very short. S_2 in males of Gomphidae, Aeschnidae, and Corduliidae has lateral oreillettes (Fig. 20).

The genital opening of the males is situated on the ventrum of S_9; in females between S_8 and S_9. In males the genital aperture is flanked by two valves, and the penis is rudimentary. In females, the area around the genital pore may be simple, or carry an ovipositor (see below). Unique to the order is the fact that the male copulatory and accessory genital organs are not situated on S_9, but on S_2, S_3, or both, and have no internal or external connection with the genital duct and the reduced peneal organ. They are secondary formations, and for that reason it is incorrect to use the terms penis or phallus for the male intromittent organ. A basic study of the morphology of the male genitalia was performed by Schmidt (1915). It was revised, completed and reinterpreted in terms of functional morphology and phylogeny by Pfau (1970, 1971). The description offered hereafter is based on the work of these two authors.

In Zygoptera (Figs. 11a, 16–17), the copulatory organ consists of a two-segmented ligula (L_1 and L_2; L_2 is sometimes called the "glans"), and evagination of the second sternite. It lies in a deep fossa of the floor of S_2 and has dorsally an open sperm groove over most of the length of L_1, the "stem". However, sperm — which are actively transferred here by the male prior to copulation — are stored in a reservoir, the vesica spermalis (VS), which belongs to S_3, and is not connected to the ligula. The reservoir opens cranially via a filling hole (MR). Further differentiations of the sternite of S_2 are the lamina anterior (LA), with a pair of outgrowths (apodemes: Ap_2), eventually

17

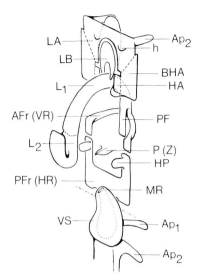

Fig. 16: Scheme of ligula and adjacent structures on segments 2 and 3
of the abdomen in Zygoptera (after Pfau, 1971)
AFr (=VR) – anterior supporting frame of hamuli;
$Ap_{1,2}$ – apodemes; BHA – base of anterior hamulus; h – hook;
HA – hamulus anterior; HP – hamulus posterior; $L_{1,2}$ – segments of ligula;
LA – lamina anterior; LB – lamina batilliformis; MR – porus of vesica
spermalis; P (=Z) – plug; PF – processus furculiformis;
PFr (=HR) – posterior frame; VS – vesica spermalis

hooks (h), on which laterally rest a pair of hamuli anteriores (HA, or ham. ant.). The latter are anchoring organs during copulation. They consist of an apical part, the hamulus proper, and a strong base (BHA). In Zygoptera, an inner thickening or plate (IP) is found. The lamina anterior shows a median invagination in which, in Zygoptera, a U-shaped lamina batilliformis (LB) finds a place. It forms a frontal protection to the base of the ligula. The ligula is fixed, by the processus furculiformis, on a sclerotized anterior frame (AFr). There is also a posterior frame (PFr) that supports a second pair of anchoring structures, the hamuli posteriores (HP or ham. post.). Medially, and between AFr and PFr, a small median lobe is found, the plug (P). Sternite 3, apart from the vesica spermalis and two pairs of apophyses (apodemes), shows no further differentiations.

In Anisoptera, the intromittent organ is not the ligula but the vesica spermalis, which has become 4-jointed (Figs. 18–19). The ligula here supports and protects the vesica (Figs. 20–21), and assists during copulation, but no longer has any copulatory function itself. The vesica has a filling pore between the 2nd and 3rd segment, against which the rudimentary penis is pressed during sperm transfer. It leads into a sperm resevoir (r). During copula, sperm flows back from it and is ejaculated into the vagina. In Aeschnidae, there is an open groove between the filling pore and the apex, a zygopteroid character. In all other Anisoptera, at least part of the groove is closed to

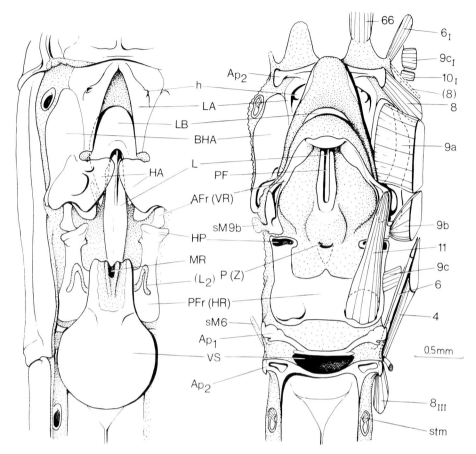

Fig. 17: Male accessory genitalia in *Calopteryx* (after Pfau, 1971)

stm – stigma

Others symbols as for Fig. 16. Numbers on right side of the figure refer to
the musculature of the accessory genitalia

form a sperm tube which, in the most evolved forms (Libellulidae), ends in an
ejaculation chamber. Swelling bodies are present in the various segments of the vesica
(Fig. 22). These are also found in the zygopteran ligula (but *not* in the anisopteran
ligula). They anchor the intromittent organ into the vagina.

The apical segment (L_2, V_4 or "glans") may bear various appendages, such as flanges
and long spines and horn-like processes. At least in *Calopteryx*, Waage (1979) has
convincingly shown that these serve to displace sperm from previous copulations
from the bursa copulatrix and spermathecal arms of the female, prior to ejaculation.

The complexity of the glans in some libellulids, involving flagellae, a sclerotized apical
outgrowth (the retinaculum, cf. Pinhey, 1970), and in- and outfolding flanges, suggests
the possibility for complex interactions of structure and function here as well, but
virtually no research has been carried out in this domain.

19

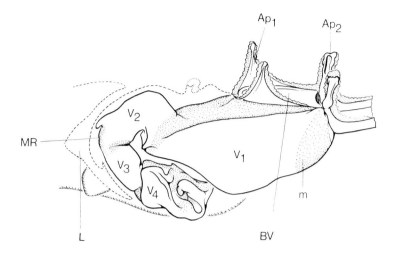

Fig. 18: Structure and position of the vesica seminalis in *Aeshna cyanea* (Müller)
(after Pfau, 1970)

$Ap_{1,2}$ – apodemes; BV – base of vesica spermalis; L – ligula;

m – caudal membranaceous part of vesica spermalis;

MR – opening (in Aeschnids: tip of slit-shaped aperture) of vesica spermalis;

V_{1-4} – segments of vesica spermalis

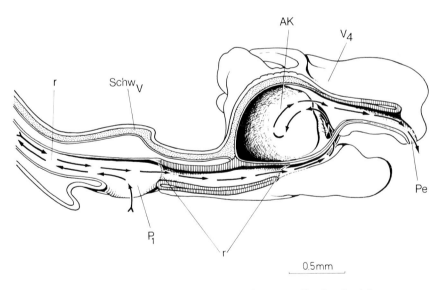

Fig. 19: Longitudinal section through the vesica spermalis of a *Cordulegaster*
(after Pfau, 1970)

AK – ejaculation chamber; P_1 – filling pore; Pe – ejaculatory pore;

r – sperm reservoir; $Schw_v$ – erectile tissue

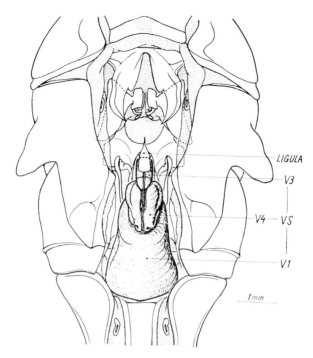

Fig. 20: Male accessory genitalia of *Aeshna cyanea* (Müller)
(after Pfau, 1970)
V_{1-4} – segments of vesica spermalis;
VS – vesica spermalis

In Anisoptera, the lamina batilliformis is reduced or absent. In many Anisoptera there are two pairs of hamuli. However, only one pair is found in the most evolved families (the Libelluloidea).

Each hamulus may then be differentiated into an inner and outer branch. In libelluloids the hind ventral angle of S_2 is produced into a hump, the lobus genitalis (lob. gen.). In Aeschnidae, the ventral margin of pleurite 2 may be variously modified.

The female genitalia are situated on the ventrum of S_8 and S_9. In all Zygoptera and in the Aeschnidae the genital opening is flanked by three pairs of gonapophyses, differentiated into valves. Together they form an ovipositor adapted to insert eggs one by one into plant tissue (endophytic mode of oviposition). In Zygoptera, the floor of S_8 is usually raised posteriorly, sometimes produced into a point (vulvar spine), and with a distal valvifer (Vf) that supports the first pair of valves (v_1). The second and third pairs sit directly on S_9. The second (v_2) is narrow, and both v_1 and v_2 are accommodated within v_3, which has a broad base (almost as broad as the length of S_9) and bears an apical pair of styli (St). Valves $v_1 + v_2$ form the perforating apparatus proper (Fig. 23), and have to be lifted out of v_3 (the latter often serrated along its margin) during oviposition. In the Cordulegasteridae, the ovipositor is long (stretching out behind the abdominal tip), and strongly built, but composed of only

21

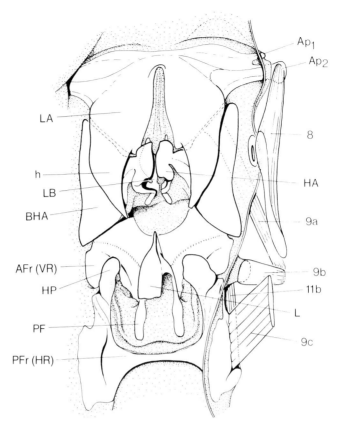

Fig. 21: Male accessory genitalia of *Aeshna cyanea* (Müller)
(vesica spermalis removed)
(after Pfau, 1971)
The supporting musculature is numbered on the right hand side of the figure
AFr (=VR) – anterior frame; Ap$_{1,2}$ – apodemes; BHA – base of anterior hamulus;
h – hook on LA; HA – hamulus anterior; HP – hamulus posterior;
L – ligula; LA – lamina anterior; LB – lamina batilliformis;
PF – processus furculiformis; PFr (=HR) – posterior frame

two pairs of valves (Fig. 24) and unfit for inserting eggs into plant material. Instead it is adapted to pushing eggs into a wet sandy or muddy bottom. In all other Anisoptera, the gonapophyses are reduced to a single pair of very small tubercles at the most (Fig. 25). The genital opening is here flanked by a pair of vulvar lips (valvulae vulvae: V.v.), of varying development.

The V.v. overlie the vulvar opening that leads into the vagina, and further into a bursa copulatrix, provided with lateral horns — the spermathecal arms (Fig. 26). Sclerites sometimes flank these structures. In *Trithemis*, these were termed sterigmata by Pinhey (1970), in analogy with other insect groups. Very few studies have as yet been conducted to elucidate the function and taxonomical value of these structures.

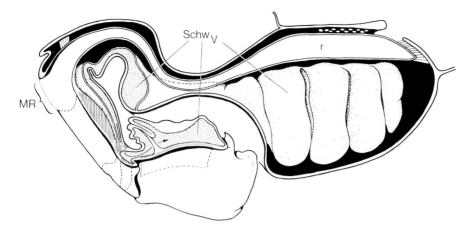

Fig. 22: Longitudinal section through vesica spermalis of *Aeshna cyanea* (Müller)
(after Pfau, 1970)
MR – opening of reservoir of vesica spermalis; r – sperm reservoir; Schw$_v$ – erectile tissue

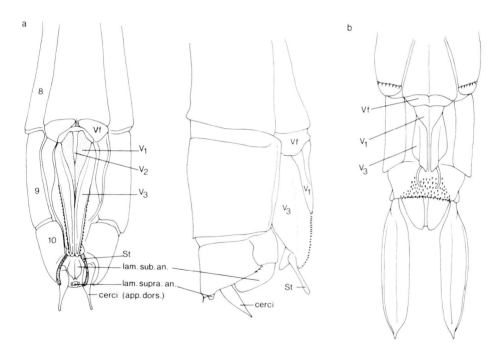

Fig. 23: Female ovipositor structure
a. Ovipositor of a zygopteran
lam. sub. an. – lamina subanalis; lam. supra an. – lamina supra-analis;
cerci (appendix dorsalis); St – styli; Vf – valvifer; V$_{1-3}$ – first to third pair of valves
8–10 – segment numbers
b. Ovipositor in an aeschnid. Only valves 1 and 3 visible

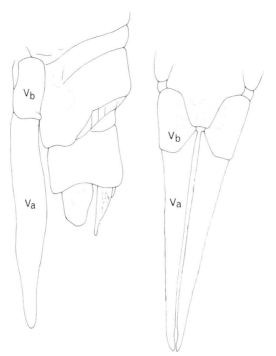

Fig. 24: Ovipositor in *Cordulegaster*
V_b – basal valves; V_a – apical valves;

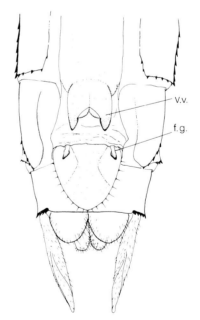

Fig. 25: Terminalia of a female libellulid *(Brachythemis fuscopalliata)*
f.g. – female gonapophyses; V.v – vulvar scales or lips (valvulae vulvae)

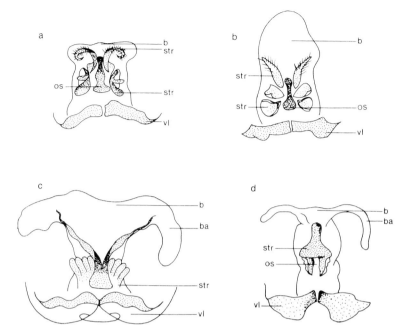

Fig. 26: Bursa copulatrix in females of *Trithemis*
a–b. *T. arteriosa*; c. *T. annulata*; d. *T. kirbyi*
b – bursa; ba – bursal arms; os – ostium (opening of bursa);
str – sterigmata (sclerotized plates); vl – vulvar lips

Anal Appendages: The anus is situated in terminal position on S_{10}. In Zygoptera, it is hidden under a rudiment of the 11th segment, the appendix dorsalis or supra-anal lamina (lam. supra an.) (Fig. 27). The mid-dorsum of S_{10} is often raised into a hump, and its posterior margin hollowed out to accommodate the lam. supra an. In Anisoptera, this lamina is greatly expanded, to form the appendix inferior (app. inf.), of very variable shape, sometimes deeply cleft, but never paired (Figs. 28–29).

In Zygoptera, two tubercles (laminae subanales, lam. sub. an.), flanking the anus ventro-laterally, and also remnants of S_{11} have grown out to form the inferior appendages. These are, consequently, always paired in this suborder, and situated below the anal opening. In all Anisoptera the lam. sub. an. are reduced to a pair of inconspicuous tubercles.

In both suborders, the superior appendages (app. sup.) or cerci are implanted terminally on the dorso-lateral angles of S_{10}. They are always paired. The specificity of the complex of appendages, laminae, circum-anal floor of S_{10}, and hind rim of S_{10} provides a remarkable example of reproductive isolation by mechanical means. The area of pronotum-lamina mesostigmalis-anterior carinal frame in female Zygoptera, or dorsum and rear of head in Anisoptera, indeed interacts as a lock-antilock system with the male terminalia.

25

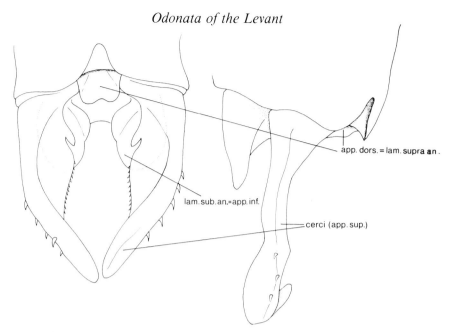

Fig. 27: Terminalia of a male zygopteran (*Lestes*), dorsal and lateral view
app. dors – appendix dorsalis or lamina supra analis; cerci – appendices superiores;
lam. sub. an. – lamina subanalis (=appendix inferior)

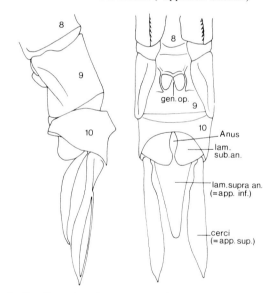

Fig. 28: Terminalia of a male anisopteran (*Aeshna*), lateral and ventral view
Anus – anal opening; cerci – appendices superiores; gen.op. – genital opening;
lam. sub. an. – lamina subanalis (these do in fact, flank the anus, and are homologous
to the app. inf. of the Zygoptera); lam. supra an. (=lamina supra analis);
these are the appendices inferiores of the Anisoptera, homologous to the appendix
dorsalis of the Zygoptera, but not to the appendices inferiores of the Zygoptera);
8–10 – segment numbers

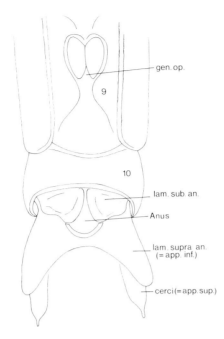

Fig. 29: Terminalia of a male gomphid (ventral view)
Symbols as for Fig. 28

NOTES ON BIOLOGY OF ADULTS

For a detailed account on the ecology of the Odonata, both larval and adult stages, the reader is referred to the book by Corbet (1962), supplemented by a recent review article (Corbet, 1980). A worked-up version of Corbet (1962) has been announced for the near future. For the ethology of Calopterygidae, Heymer (1973) may be consulted, and for *Epallage fatime* — Heymer (1975).

I here present in brief some essential features of dragonfly behaviour, partly in conjunction with what has been said earlier about (functional) morphology.

EGGS

There are two modes of egg-laying. In all Zygoptera and in the Aeschnidae, the female inserts eggs one by one in living or dead plant tissues. She may herein be accompanied by the male in tandem formation, or the male may perform a patrol flight in the neighbourhood of the female, or the female may oviposit solitarily. Oviposition may occur in branches and twigs of trees overhanging or adjacent to water (*Lestes viridis* selects willow and oak), or into submerged vegetation. Here too, there may be selective or aselective oviposition (the European *Aeshna viridis* selects *Stratiotes aloides*; *Pseudagrion syriacum* possibly selects *Mentha* sp.) and the female may partially descend into the water while ovipositing. A variant of this endophytic mode of

27

oviposition is found in *Cordulegaster*, where the female always oviposits alone in shallow marginal areas of running waters or springs. She pushes her eggs into the bottom sand or mud, hovering in a vertical position above the water surface, and bouncing up and down at a rate of about 30–40 egg-layings per minute. All endophytically laid eggs are fusiform.

In the Gomphidae and the Libelluloidea, egg-laying is exophytic. Although the female will sometimes oviposit from a perched position, and release whole strings of eggs at a time into the water, eggs are usually released one by one or in packages by airborne females. In some females (*Sympetrum*, *Crocothemis*), where the vulvar lips or scales are produced into a scoop-like structure, eggs are not dropped but projected into the water with considerable speed and accuracy. Occasionally, egg-laying occurs on the dry beds of temporary pools or rivers. Exophytically laid eggs are globular or subglobular.

LARVAL LIFE AND EMERGENCE OF THE IMAGO

Almost all dragonfly larvae are aquatic. From the egg hatches a prolarva, which quickly moults into a first instar larva. After a series of larval moults, and a period of time ranging from six weeks (in some *Ischnura*) to five years (in some *Cordulegaster*), the final instar larva leaves the water and the adult emerges from it. Emergence occurs on stems of emergent aquatic vegetation, or on rocks, or on whatever more or less vertically oriented substrate available. It begins with a fissure of the larval skin on the mid-dorsum of the thorax; synthorax, wings and head are first retracted from the larval skin. The legs unfold and immediately grasp the substratum (Zygoptera), or the anterior half of the body temporarily bends over backwards first (Anisoptera). The caudal half of the abdomen is then pulled out of the exuvia, and the wings, later the abdomen, are stretched through pumping movements. Emergent dragonflies are very vulnerable to predation. Visual predation by birds may be reduced by emergence either at dawn or at dusk, but this is no absolute rule, and there is also a heavy loss to non-visual predators such as spiders and small insectivorous mammals.

Recently, I have reported circumstantial evidence for estivation in the larvae of two libellulids that are widespread in desert environments, *Orthetrum chrysostigma* and *Trithemis arteriosa*. Larvae were found buried in damp sand at depths of about 30 cm, in beds of temporary pools in the Sahara (Dumont, 1979). Estivation is evidently adaptive in areas where rainfall is erratic.

All dragonfly larvae are carnivorous. According to their size, they feed on zooplankton, insect larvae, or even fish fry and small fish. Most larvae of Zygoptera live among aquatic vegetation, as do those of Aeschnidae and many Libellulidae. Some Libellulidae, *Cordulegaster* and most gomphids spend their larval life in mud or sand, with only the eyes and antennae emerging. Among the regional species, *Epallage fatime*, *Calopteryx syriaca*, *Pseudagrion syriacum*, all gomphids (except *Paragomphus sinaiticus*), *Cordulegaster insignis*, *Caliaeschna microstigma*, and *Zygonyx torrida* are rheophilous. All other species occur in stagnant water as well.

28

Prey are captured by a highly specialized labium, modified into a two-jointed "mask". The two segments, connected by a hinge joint, are folded back under the head when at rest, but can be extended at great speed when a suitable prey comes within reach. Details on the morphology of the mask, which is of great taxonomical importance, are given in the section on larval identification.

ADULT LIFE: MATING STRATEGIES

After a period of maturation — often spent away from water — during which first the wings dry out completely, the body colours fully develop and, internally, the gonads ripen, males return to water. Females continue to spend much time away from water, and usually return for mating and oviposition only. Males are often territorial. Territories may be ephemeral or permanent, and need to be re-selected, re-established, and defended against competitors daily. Territories are primarily sites where courtship display and mating can take place. In some cases, they are oviposition sanctuaries for females, and hunting grounds for males.

Courtship display has been intensely studied in *Platycnemis* and *Calopteryx* (Buccholz, 1955, 1956; Heymer, 1973, 1975). In both genera, pre-contact recognition of a conspecific male by the female takes place. This involves a characteristic dance by the male, in which the shape and colour of legs *(Platycnemis)*, and colour of the ventrum of the male's terminal abdominal segments are exhibited to the female. In *Calopteryx* males, extensive wing clapping produces a distinct stroboscopic effect due to the contrasting hyaline and coloured portions of the wings, which also contributes to female acceptance or refusal of tandem formation. This ethological mechanism of reproductive isolation is so strong that the structural (or mechanical) aspect has lost all selective pressure: the male terminalia and female pronotal structure in all West-Palaearctic species of *Calopteryx* are generalized, and thus selectively neutral. In *Platycnemis*, this is not absolutely true: the leg display has not eliminated mechanical reproductive isolation completely, and structural differences between the key-areas in both sexes are still present. In other coenagrionids, the lock-antilock mechanism prevails although, as shown by many authors, females may still visually pre-select males on sighting certain areas of the body, usually the terminal abdominal segments that show characteristically coloured areas.

There are also many tactile setae on the pronotum of the female, and on the superior appendages of the male (see figures in the text), which suggest that fine mechanoreceptive stimuli in both sexes might be releasers of further copulatory display. Tennessen (1975) is of the opinion that chemoreception may play some role, although there is no conclusive evidence for this hypothesis at present.

Males land on the dorsum of the female synthorax, quickly run forwards, and grasp the female's pronotum and lamina mesostigmalis (in Zygoptera) with their terminalia. In Anisoptera, courtship display is reduced or absent, and males grasp females across the vertex and the frons with their inferior appendages, while the superior appendages hold the rear of the head, eventually the anterior collar of the pronotum. In Anisoptera, it is often possible to ascertain whether females have copulated or not:

29

the male inferior appendix frequently leaves copulation scars on the compound eyes, by damaging the ommatidia that lie immediately adjacent to the vertex.

Once a successful tandem has been formed, and sometimes even prior thereto, males have to "charge" their vesica spermalis with sperm. To do this, they curve the abdomen backwards, and press the rudimentary penis against the filling pore of the vesica. The male then invites the female, mainly by lifting her gently, to press her genitalia against his secondary copulatory complex (formation of a copulatory "wheel"); anchorage, at this stage, is mainly achieved by the hamuli. The ligula (Zygoptera only) is then moved caudad into the vagina, where it anchors itself by means of the swelling bodies. Sperm is released along the dorsal groove in the ligula by a downward movement of S_3, which compresses the caudal area of the vesica and pushes the sperm mass into a "genital aperture" formed by the female vulva and the ligular groove. The latter is closed up (so that no losses of sperm occur) by the plug (P).

As indicated above, the appendages of ligular segment 2 (the "glans"), at least in *Calopteryx*, but probably in many genera and species, fulfill another function as well: prior to sperm transfer, they empty the bursal arms and the bursa copulatrix of the female of sperm deposited there during previous matings. Since these apical flanges are widespread in Zygoptera, and copulation is here a rather long event, this behaviour is possibly of widespread occurrence (perhaps also in Aeschnidae, Gomphidae?).

In many libelluloids, tandem formation and mating are extremely quick events that may take place during flight, and oviposition often occurs immediately after copulation. It is therefore not clear whether the vesica spermalis plays a dual role here as well, in spite of the elaborate structures that are found on the glans. It has been suggested that sperm from previous matings may be pushed beyond the level where the oviducts reach the bursa, so that "old" sperm is no longer available for fertilizing eggs that move down the oviducts.

In Aeschnidae, the vesica spermalis, supported by the ligula, is introduced into the vagina, and anchors itself by means of its swelling bodies (no such structures are found in the ligula here). Sperm is transferred through the open grooves on the third segment of the vesica, by compression of the basal sperm reservoir. The injection pore is situated between S_3 and S_4. This is an archaic system, and suggests that the Aeschnoidea are monophyletic and plesiomorphic to all other Anisoptera. The top segment of the vesica has no role in sperm transfer: it unfolds some of its apical appendages for better anchoring (and for removal of sperm from previous matings?).

In all other Anisoptera, the third segment has a closed sperm tube, and an ejaculation chamber in the fourth segment, that acts as a suction-and-pressure pump.

COLORATION

Bright metallic cuticular colours are found in the adults of the Calopterygidae and in *Lestes* where they extend over the whole body and, in *Calopteryx*, even on the main wing veins. In the latter genus there is sexual dimorphism in both colours and coloured wing spots, males usually being blue and females green. Metallic colours are also found in a few libellulids (*Rhyothemis, Zygonyx*), but are often restricted to small areas of the body, such as the frons in some species of *Trithemis.*

All other Odonata have subcuticular lipoprotein colours, with blues often dominant in males and greens in females, but numerous other colours (often tinges of yellow, brown and red) are also frequent. In quite a few species, a waxy blue, blue-grey or white pruinosity may develop secondarily on thorax and abdomen as specimens age.

DISTRIBUTION OF THE ORDER

Odonata occur worldwide. The total number of species may be estimated at between 4,000 and 6,000. The vast majority of these occur in tropical and subtropical climate zones. The West-Palaearctic dragonfly fauna has suffered much from Pleistocene glaciations and, today, is greatly impoverished, numbering some 220 species. The reason for this is that, unlike in North America, there were only limited possibilities for refugia to form, and, after the glaciations, recolonization from other areas was hampered by desert and mountain barriers. The two most important refugia were the Maghrebo-Iberian and the Anatolian-Levantine ones.

BIOGEOGRAPHICAL COMPOSITION OF THE LEVANTINE DRAGONFLY FAUNA

The 82 species included within the Levantine fauna (and Sinai) represent very different affinities, as could have been expected from the general botanical and zoological knowledge of the area. Bodenheimer (1937, 1938), Schmidt (1938), and more recently Por (1975) have outlined the complexity and the richness of the regional fauna, situated at the crossroads of the Palaearctic, Oriental and Afrotropical regions.

Biographical Nature of the Regional Fauna

I have divided all elements into 10 categories, that can be reduced to three major classes: species with restricted ranges ("endemics"), species with a wide range (i.e. the whole or greater part of a major biogeographical area), and species with very wide ranges (i.e. occurring in at least two major biogeographical areas). For the endemic species, I have indicated — wherever possible — to which of the wider ranges their nearest relative(s) belong and, finally, the relative importance of the various elements — expressed as a percentage of the whole fauna.

The categories are:

I. Restricted ranges: 1. Endemics of the Anatolian, Levantine and Mesopotamian provinces; 2. Endemics of the Levant (the Orontes, Litani and Jordan river systems); 3. Endemics of the Jordan Valley.

II. Wide ranges: 4. Saharian (including the Arabian desert); 5. Afrotropical; 6. Oriental; 7. South-Mesasiatic (the steppe Irano-Turanian province of Asia); 8. West-Palaearctic.

III. Very wide ranges: 9. Combined Afrotropical and Oriental areas; 10. Combined Afrotropical, Oriental and Palaearctic areas.

The first remarkable fact is the high percentage of endemic species (about one third of the total). About one-fourth of the fauna is restricted to the Eastern Mediterranean basin, while some 9% are endemic to only a small part of that area. If one tracks down the phylogenetic origin of these endemics, 22% appear to be derived from a Palaearctic ancestry, only 6% are from African stock, and none is from truly Oriental stock.

Among the wide-ranging species, the situation is more balanced: 19% of the species are either of West-Palaearctic or African origin, while 17% are Asiatic. The very wide-ranging, often subcosmopolitan species are remarkably few in number, totalling only 10%.

Summing up all species (that is, both endemics and wide ranging elements) according to the respective faunal areas where they belong or stem from, the predominance of the West-Palaearctic element is seen to be overwhelming: more than 40%. The Afrotropical element comes next (25%), while the Asiatic element represents only 17% (and truly Oriental species not more than 5%) of the fauna.

There are steep north–south-oriented gradients in the dragonfly fauna of the Levant. Roughly, north of Por's (1975) Nehring line, one finds more Palaearctic species, while south of the Bodenheimer line, the number of Saharian and Irano-Turanian elements increases sharply. However, the area also holds two important relict "pockets", mainly for Afrotropical species. One is the former Lake H̠ula where, until its recent drainage (but probably much more as a sequel to the eutrophication of the remaining lakelet), at least three endemic subspecies of African species existed. Only one is extant today. The second one is the Dead Sea basin, and primarily its eastern side, where several permanent rivers (the largest one being Wadi Mujib or Nah̠al Arnon) drain into the Dead Sea.

The composite fauna of the Levant is the outcome of a series of historical events, such as the absence of glaciation during the Pleistocene, that made the Anatolian-Levantine zone a major focus for the survival of a Palaearctic faunal segment destroyed elsewhere. That there was no glaciation does not mean that no climatic fluctuations occurred. They were expressed in temporarily increased or decreased precipitation and mean annual temperature. Probably, the maximum of climate change did not go in phase with the glacial maximum in Europe and the Pontic Alps. This means that maximum faunal interchange between the Oriental and Afrotropical areas, partly via the Levant, occurred either before or after the glacial maximum.

No fossil evidence of these events is available for the Odonata. Therefore, unless speciation occurred *and* the species that evolved could survive the later influx of competitors, repetitive migrant waves from both regions involved must have introgressed with previously established populations. Thus, little can be said about immigrant waves much older than the latest main stadial of the Würm, that peaked at about 20,000–18,000 BP. Shortly before that stadial, a major warm pluvial epoch bridged the Sahara (and Sinai) to the extent of creating permanent running waters other than the Nile. This permitted the ancestor of rheophilous species such as *Pseudagrion syriacum* (closely linked to *Mentha*, growing in permanently flowing rivulets), and the ancestor of a slowly dispersing species such as *Agriocnemis sania* (which may, in fact, have been the Oriental *A. pygmaea* Rambur, today still occurring on the Seychelles), to extend their ranges. During the subsequent desiccation of the Sahara (and in fact of most of Africa) they evolved into their present status. Whether the endemics of Lake Hula underwent the same period of isolation from their African stock is unknown but not improbable. From a more general study of the aquatic biota of northern Africa (Dumont, 1979), it indeed appears that at least 20,000 years are needed to furnish evidence of allopatric speciation. In Sahara, *Ischnura saharensis* (Aguesse), a sister-taxon to the insular (Sardinia, Corsica, Sicily) *I. genei* Rambur, is an example. This species has a range that covers the entire Sahara Desert. In the east, it is limited by the Nile Valley, where *I. senegalensis* occurs. Both species are mutually exclusive, and *I. senegalensis* has managed to spread into Sinai and as far as the Dead Sea depression. Here, it meets with another member of the *I. elegans*-group *(I.e. ebneri)*, and again, both show little or no intergrading.

Clearly, many more species than presently extant must have made intercontinental crossings during this Late Pleistocene humid phase. The Oriental anisopteran *Hemicordulia asiatica* Sélys, which was first found in Africa on Lake Victoria (Pinhey, 1961), and later in other localities of central East Africa as well, may possibly have reached that area at this time, but could not maintain colonies in the vast intermediate zone between India and East Africa in later periods. There may have been many cases of this nature. One important fact, not often stressed in biogeographical reasoning, is that not all waves of immigrant species are necessarily successful in establishing themselves in newly conquered areas in the long run, even in the absence of major environmental changes. Immigrants indeed face a pre-existing fauna, with which they have to compete for niches. Slight environmental changes in their new biotopes, insufficient to eliminate the species if it were occurring there alone, may still tip the balance of competition in favour of previously established species, and thus ultimately cause the extinction of the apparently successful invaders. By this type of interaction, one may intuitively grasp the reasons behind the curious fact that of the three *Pseudagrion* species occuring in the Levant, only one is found in the Nile delta, while only one of the three deltaic *Pseudagrion* of Egypt is also known to occur in the Jordan Valley.

The latest glacial peak, around 20,000 BP, causing great aridity and a lowered mean annual temperature over most of Africa, created unfavourable conditions for tropical dragonflies in the Levant, and certainly eliminated many. One strong argument for this is that, through a eustatic 120–130 m lowering of the sea level, close contact between Corsardinia and Tunisia on the one hand, and Italo-Sicily and Tunisia on the other hand, was established. Relatively few dragonfly species (but still a substantial number) crossed over on this occasion. None of them were tropical in nature, and very few were Oriental or Irano-Turanian, indicating that the route from Asia via the Levant was not a very effective pathway at that time. However, the Hula and Dead Sea pockets were probably fully exploited in preserving relict populations.

Following warming up of the climate and deglaciation, a new wave of north–south migration started. It is quite conceivable that at least part of the Afrotropical fauna

33

of the Levant re-immigrated into the area during the post-Würm humid spell of 12,0000–7,000 BP, and that the Nile was its major pathway. Quite a few of the South-Mesasiatic species in the Levantine fauna may also have reappeared here at that point in time. Crossing Sinai, and moving into North Africa, they did indeed penetrate rather deeply into the Sahara (sometimes even crossing it). Some reached the Iberian Peninsula, but none could colonize Sardinia, Corsica, or Sicily which, as a result of the rising of the sea level that accompanied the melting of the ice caps in Europe, had become out of reach for any colonizer lacking strong migratory powers.

It should be added that there has also been faunal exchange between Africa and Asia via the south of the Arabian Peninsula, with species crossing the Straits of Hormuz and Bab el Mandeb.

The dragonfly fauna of Saudi Arabia, as far as is known, mainly consists of two segments: Mesasiatic species, and Afrotropical species. The Palaearctic element is negligible (Waterston, 1980). The fauna of the Nile Valley, likewise, is dominated by Afrotropical species, a broad segment of Mesasiatic species, and only very few Palaearctic "relics" (e.g., *Sympecma fusca*) (Dumont, 1980).

COLLECTION AND PRESERVATION

Adults may be collected with a butterfly net, and killed in a killing-jar using cyanide, ether, or any quick-killing agent. Species with metallic body colours can be preserved without further manipulation, and stored in collection either pinned or in papers. Pruinescent species and dull brown, ochre or olivaceous species may also be preserved directly, and will not show drastic colour changes post mortem. Most coenagrionids and aeschnids, however, with delicate subcutaneous blue and green colours, will turn black some time after death, unless treated with a lipid solvent. Acetone is most widely used for this purpose. Coenagrionids should be immersed completely in an acetone bath, and left there for at least a few hours. In aeschnids, especially females, the ventrum of the abdomen should be opened and the viscera removed. Care should be taken not to scratch the internal surface of the tergites, thus damaging the colour layer. With the aid of a rather blunt pin, a few holes can be punched into the ventrum of the thorax and the rear of the compound eyes, in order to allow better impregnation by the solvent. However, in spite of this, the deep bright colours of the eyes will not normally preserve well, even if carefully treated. Aeschnids should not be left in acetone for more than, say, half an hour. Acetone is, indeed, a powerful water extractant as well, and animals treated with it will tend to become very brittle. This can be partly overcome by the addition of a few drops of glycerol (as is recommended by those who use alcohols).

If it is necessary to preserve animals in liquid permanently (which may be desirable if at a later stage anatomical observations have to be made), alcohol (either ethyl alcohol or a higher homologue) should be used, never acetone.

A less widely used, though quite productive way of obtaining dragonflies, is to collect live larvae, breed them in aquaria, and allow adults to emerge from them. In the case of powerful flyers such as most aeschnids, this is often the easier way of securing specimens. A special advantage, in particular in an area like The Levant where so many species are not yet known in the larval stage, is that the exuvia can be equated directly with the adults emerging from them. Freshly emerged specimens should not be killed immediately, but allowed to maturate in cages for a few days (no feeding is normally required), until their cuticle has hardened and their body colours are fully developed.

SYSTEMATIC PART

Order ODONATA

Key to the Suborders of Odonata

1. Both pairs of wings of similar shape; membranula absent. Labrum with glossae of the same shape as paraglossae, mostly deeply cleft in the middle. d without a ht.

    Compound eyes widely separated, with ocelli implanted between them on a flat, undifferentiated vertex, and arranged in a triangle.

    Males with superior and inferior appendages paired, and the inferior appendix implanted below the anus.

    Vesica spermalis simple, not segmented. Lamina batilliformis well developed. Females with an ovipositor.

    Abdomen always cylindrical.   **Zygoptera**

\- Hind wing with a broader base than forewing, and usually with a membranula. d with a ht.

    Glossa usually smaller than paraglossae and fused into a single ligula (except in *Cordulegaster*). Compound eyes confluent in a point or over a longer distance; if separated, vertex usually with differentiations and ocelli more or less lined.

    Males with paired superior appendages, but only one inferior appendix, although it may be deeply cleft. Appendix inferior situated above the anus. Vesica spermalis strongly developed, 4-segmented.

    Females with or without ovipositor.

    Abdomen cylindrical or depressed.   **Anisoptera**

Suborder ZYGOPTERA

Key to the Families of Zygoptera

1. Numerous an; wings not petiolated. Analis not confluent with hind wing margin at wing base   2

\- Two an; wings petiolated. Analis forms hind margin of wings between base and Ac   3

2. Lateral suture between mesothorax and metathorax complete. Discoidal cell elongate, traversed. A pseudopterostigma in the females, no Pt at all in the male. Body brilliant metallic blue or green.   **Calopterygidae**

\- Lateral suture between mesothorax and metathorax incomplete. Discoidal cell simple, not traversed. Body not metallic. Pt present in both sexes.   **Euphaeidae**

3.   IR$_3$ and R$_{4+5}$ arise closer to level of arc than to level of N. Pt elongate, almost as long as the two underlying cells. Body often green metallic.   **Lestidae**

–   IR$_3$ and R$_{4+5}$ arise almost directly below N. Pt less than twice as long as wide, and much shorter than the combined length of the two underlying cells (measured along R$_1$)   4

4.   d quadrangular. Tibiae of legs 2 and 3 often greatly expanded.   **Platycnemididae**

–   d a trapezium. Tibiae of legs 2 and 3 never expanded.   **Coenagrionidae**

Family CALOPTERYGIDAE

Medium- to large-sized damselflies, with first lateral suture on synthorax complete and non-petiolated wings. Wing venation very dense, with numerous antenodal and postnodal cross-veins. Anal vein separated from hind wing margin along its entire length; an anal field, composed of rather numerous cells, is present. No true pterostigma in either sex. A pseudopterostigma (traversed by cross-veins) in the females. d many times longer than wide, traversed by numerous cross veins. Wings with or without blue coloured spot.
One regional genus.

Genus CALOPTERYX Leach, 1815 (emend. Burmeister, 1839:825)
Edinburgh Encyclopaedia, 9:137. *(Calepteryx)*

Agrion Fabricius, 1775:425.

Type Species: *Libellula virgo* Linnaeus, 1758.
Medium-sized damselflies, with blue or green metallic body sheens. Female with or without white pseudopterostigma. App. sup. of males not divided, simple, forcipate, slightly longer than S$_{10}$. App. inf. cylindrical, straight, about 2/3 the length of the app. sup.
Restricted to running water, *Calopteryx* species are poor dispersers. With time, inhabitants of single stream basins therefore develop recognizable population characters which in many cases, however, have not reached the level of "official" taxonomical ranks.
One relatively well-defined species inhabits the Jordan Rift Valley. A possible second species is found on the Litani and extends well into Syria, where a third taxon, believed to be part of the *C. splendens*-group, is also found. The genus does not extend to Sinai and does not occur in the lower Nile Valley.
Distribution: North America, Europe, North and Central Asia, including Japan.

Calopterygidae: Calopteryx

Key to the Species of Calopteryx
(Figs. 6, 9, 14, 17, 30–31)

1. Body colours metallic green in both sexes. No Pt or pseudo-Pt. No coloured wing spots in either sex. Male with ventrum of S_{8-10} greenish-ochraceous.   **C. hyalina** (Martin)
 – Body colours metallic blue in the males, and a coloured wing spot always present on all wings. Ventrum of S_{8-10} bright ochraceous. Females with or without coloured wing spots
 2

2. Males with wing spot extending between apex of wing and nodus, usually reaching well basal to nodus. Females without (heterochrome) or with (homochrome) wing spots on all wings. Legs black.   **C. splendens intermedia** Selys
 – Males with wing spot extending between apex of wing and half to 2/3 the distance between apex and nodus, never reaching the nodus. Females usually heterochrome.
 Pseudo-homochromes have the apex of the *hind wing* enfumed or darkened. Flexor side of all femora lightly coloured (covered by bluish pruinosity in mature specimens).

 C. syriaca Rambur

Calopteryx syriaca Rambur, 1842
Figs. 30a, b, 31

Calopteryx syriaca Rambur, 1842:223. Selys & Hagen, 1854:32; Dumont, 1977b:131.
Calopteryx syriaca syriaca —. Bartenef, 1912:69.
Calopteryx splendens syriaca —. Morton, 1924:27; Schmidt, 1938:136.

Type Locality: Mount Lebanon.

Male
Head: Mouth parts black, except cardo and stipes of maxilla (yellow), and labrum, which is yellow with a median black marking and black margins. Pedicellum of antennae bright yellow. Rest of head blue-green metallic, often with purple sheen; rear of head with occipital corners produced into a ridge which, in dorsal view, appears as a pair of lateral tubercles.
Thorax: Pronotum and synthorax entirely metallic. Sutures black, except Su_2 which is overlaid by a yellow band with black margins, widening over inf_3 and including the metathoracic stigma. Suture with metasternum yellow too. Legs black. Femora yellow on flexor side. In fully mature specimens, this yellow colour often masked by a whitish-blue pruinescent layer that also covers the ventrum of the thorax and the first abdominal segments.
Abdomen: Dorsum and sides entirely metallic blue. The tergites overfold the sternites, and are yellow at their margins. The sternites are black, with a spiny median carina. Some yellow at the end-rings of the segments only. This yellow colour increases in width posteriorly, and the sternites of S_{8-10} are all bright ochraceous, with a narrow median black stripe on S_8, sometimes a rudimentary one on S_9. App. inf. also

Fig. 30: *Calopteryx* spp.
a. wings of *Calopteryx syriaca* Rambur, 1842; male;
b. aged female of *C. syriaca* with enfumed hind wings;
c. *Calopteryx splendens intermedia* Selys, 1887; male (Selys' type)

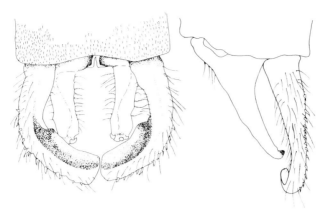

Fig. 31: *Calopteryx syriaca* Rambur, 1842
male terminalia, dorsal and lateral views

ochraceous in ventral view, with metallic tips. App. sup. metallic black. Accessory genitalia: lam. ant. (LA) with a tubercle on each side; ham. ant. (HA) black, strongly developed; ham. post. (HP) small, hollowed-out. Ligula with flanges and horn-like appendages on apical segment. Vesica elongate.

Wings: Apex with coloured spot. Never a hyaline fringe between the wing tip and the spot. Basal extent of the spot variable. It may reach fairly close to the nodus (but never attain it), or extend not further than midway between the apex and the nodus. The wing spot is already clearly visible in freshly emerged specimens.

Female

Head and thorax as in the male, but green metallic. Dorsum and sides of abdomen green, sometimes with a cupric sheen on the terminal segments. A narrow yellow stripe may be seen on the mid-dorsum of S_{9-10}. The rims of all tergites are broadly yellow. The ovipositor is entirely yellow. In older females, pruinosity may extend over all sternites of the abdomen.

Wings: R_1 distinctly curved. A bright white pseudopterostigma, traversed by up to 6 cross-veins, often smaller in hind wing than in forewing. Never a wing spot on the forewings. In some specimens the apex of the hind wing can become enfumed to the extent of developing a pseudo-wing spot. It is, however, never as opaque as in the males.

Measurements (mm): Male. Total length 47–54; abdomen 37–42. *Female.* Total length 44–48; abdomen 34–37.

Distribution: Common in Syria, the Lebanon, Jordan and Israel, on the following river systems: Asi (Orontes), Litani, Yarmouk, and Jordan. Also on the eastern drainage of the Dead Sea: Wadi Mujib, Wadi Wala, Wadi Hasa and on some of the short coastal rivers of Israel. Not extending beyond the Dead Sea. Specimens have been captured between March and October. The species is very sensitive to pollution, and stream regularization schemes also affect it greatly.

Israel (Locality records): Benot-Ya'aqov bridge on River Jordan (1), East of Lake Hula (1), Gadot (1), Gonén (1), Dan (1), Tel el Qadi (1), Sedé Nehemya (1), Nahal 'Ammud (2), Wadi Bira (2), 'Akko (4), Bet She'an (7), Tel 'Amal (7), Lake Kinneret (7), Ashdot Ya'aqov (7), Yarmouk River (7), Migdal (7), Deganya (7), Tabigha (7), Wadi Fari'a (12), Wadi Qilt (13).

Also recorded from Nahr ez Zerka, from Wadi Mujib (Nahal Arnon) on the eastern side of the Dead Sea, and from the tributaries of the Jordan River in Jordan (Wadi Yabis, Wadi ez Ziglab, Wadi Taiyiba), although damming and regularization schemes during the 1970s have greatly reduced its populations there (Schneider, pers. comm.).

Calopteryx hyalina (Martin, 1909)

Calopteryx splendens hyalina Martin, 1909:213.
Calopteryx syriaca hyalina —. Bartenef, 1912:72.
Calopteryx hyalina —. Dumont, 1977b:131.

Type Locality: Lake Homs, Syria.

Male
Structure and dimensions as in *C. syriaca*, except for the absence of a coloured wing spot, and the flexor side of the femora, which is not yellow. The tibiae may be dark brown instead of black.
Female
Hard to distinguish from females of other regional *Calopteryx* species. The legs, which are uniformly brown or black, may serve as a discriminatory character, but this is certainly not infallible. No dark spot develops on the hind wing in females. The pseudopterostigma, in all specimens I have seen, was of similar shape in both pairs of wings, but this character should be rechecked on longer series.
Distribution: *C. hyalina* was described from the area of Lake Homs. Buccholtz (1955) found this species in the coastal areas of Syria and the Lebanon, as far south as the lower Litani. Ethological isolation from *C. syriaca*, with which it co-occurs in the Levant, was observed. *C. hyalina* does not seem to cross the Jordan-Litani divide.

Calopteryx splendens intermedia Selys, 1887
Fig. 30c

Calopteryx splendens race *intermedia* Selys, 1887:39.
Calopteryx intermedia intermedia —. Bartenef, 1912:77; Schmidt, 1954b:244.
Calopteryx xanthostoma intermedia Selys, 1882 (sic!): St. Quentin, 1965:534.
Calopteryx splendens intermedia intermedia —. Bartenef, 1930:523.
Calopteryx splendens intermedia —. Buccholtz, 1955:366; Dumont, 1977b:130.

Type Locality: Ekbaz, Hatay Prov., Turkey.

42

Structurally and by dimensions, this species, too, cannot be distinguished from *C. syriaca*. Males are characterized by a large wing spot that extends between the very apex of the wing (only rarely is there a very narrow hyaline fringe on the apex) and 5–20 cells basal to the nodus. In specimens with a maximum extent of the wing spot, half of the space between the nodus and the wing base is coloured. In specimens with a minimum extent of the wing spot, this coloured area extends only slightly basal to the nodus. Both types may occur as extremes within a single population. The basal margin of the wing spot may be smoothly rounded, or deeply crenulate. Both types, again, may co-occur within single populations. Heterochrome females cannot be told apart from heterochromes of *C. syriaca*, except sometimes by the colour of the femora. Heterochrome females of *C. s. intermedia* have a deep brown wing spot, with a strongly contrasting, large, pure white pseudopterostigma. The basal extent of the wing spot is usually as in minimum-males, and the basal margin of the spot is, as a rule, smoothly rounded. The percentage occurrence of homochrome females within single populations varies greatly.

In both males and homochrome females, the wing spot is visible immediately upon emergence.

Measurements as in *C. syriaca*.

Distribution: A mountain species that occurs on running waters and limnocrenes in Anatolia, Iran and northern Iraq. It extends to the Amanus Mountains in the south (where the type locality is situated). Possibly, isolated colonies occur in the Lebanon and Anti-Lebanon ranges.

Family EUPHAEIDAE

Robustly built damselflies with large, globular eyes, and a rounded frons, which is not angulate. Prothorax with a large rounded lobe lateral to the middle lobes; posterior rim broad but simple, without medial differentiations. Synthorax robust, usually banded (in males often uniformly coloured by secondary melanism). Wings hyaline, enfumed in females, with or without apical spots, shortly or not at all petiolated. Reticulation fairly close, with many supplementary sectors between the main veins. Numerous antenodal and postnodal cross-veins; primary antenodals not indicated. Pterostigma long and narrow, with oblique ends. Abdomen cylindrical, tip of S_{10} not raised. Anal appendages of males: superiors forcipate, strongly built, as long as or longer than S_{10}. Inferiors small to moderately large, but always shorter than superiors. Females with anal appendages as long as S_{10}, acutely pointed. Ovipositor robust. Abdomen with lateral stripes.

One regional genus.

Genus EPALLAGE Charpentier, 1840

Libellulinae eur., p. 6

Type Species: *Agrion fatime* Charpentier, 1840.
The genus is monotypic.

Epallage fatime (Charpentier, 1840)

Figs. 32–35

Agrion fatime Charpentier, 1840:132.
Epallage fatime —. Schneider, 1845:115; Selys, 1853:50; Selys & Hagen, 1854:165; Fraser, 1934:76; Morton, 1924:28; Schmidt, 1938:136; Schmidt, 1954a:62; Dumont, 1977b:133.
Euphaea fatima —. Selys & Hagen, 1850:143.

Type Locality: Thracia (Turkey-in-Europe).

Male

Head: Mouth parts and genae largely ochraceous, darkening with age. Labrum ochraceous, with a median black tongue. Clypeus yellow, marked with black. Frons, vertex and occiput black, covered with a dark blue pruinosity.

Prothorax black and entirely pruinescent. In freshly emerged specimens, a colour pattern very similar to that in the female is seen, with extensive yellow on vertex and occiput, and on the anterior collar of the pronotum, and with yellow antehumeral and lateral thoracic bandings. Yellow stripes are also observed on the femora and the tibiae. Very soon, however, all these clear parts darken, and become covered by a waxy blue coating. Legs short.

Wings hyaline with brown apical fringe. The extent of this brown marking varies from a hardly visible rudiment to a spot extending halfway to the pterostigma, which is black and elongate (typically as long as five underlying cells). 11–17 an, 14–21 pn. Anal vein describing an anal loop in both pairs of wings.

Anal appendages black. App. sup. distinctly longer than S_{10}, robust. In dorsal view, a single very strong, downwardly turned apical hook is seen. In lateral view, the external apical area of the appendix also bears a second, less robust hook, and between the two hooks lies a cavity, which grasps and holds the lateral angles of the posterior lobe of the female's pronotum in tandem formation. The inner surface of the main hook is particularly smooth, and rests on the lam. mes. during copula. The app. inf. are 2/3 the length of the app. sup., conical, with a small subapical spine. Accessory genitalia: lam. ant.(LA) without tubercles; ham. ant. (HA) hammer-shaped; ham. post. (HP) small, twisted. Vesica spermalis heart-shaped, with quadrangular apex.

Female

Differs from the male in being more stoutly built, and in showing reduced pruinescence. Yellow colours are far more extensive. On the dorsum of the head, only the areas around the ocelli and the suture between frons and vertex are black in young

44

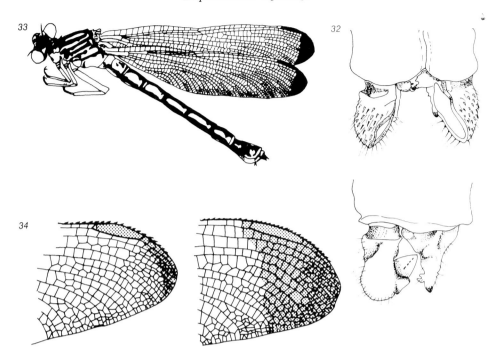

Figs. 32–34: *Epallage fatime* (Charpentier, 1840)
32. male terminalia, dorsal and lateral views;
33. habitus and markings of a teneral female;
34. minimum and maximum extent of apical wing spots

females. Gradually, black colours develop, so that finally only isolated yellow spots beside the ocelli and on the occiput remain. Similarly, the pronotum darkens considerably with time, except for the lateral bosses. The ground colour of the synthorax and the abdomen, likewise, is yellow. The carina is flanked by two black stripes and by a pair of black humeral bands. There is a narrow black humeral stripe, extensive black markings on Su_1 and Su_2, and black patches on epm_2, epm_3, and on inf_2 and inf_3. The femora of all pairs of legs also show longitudinal yellow stripes.

The wings have apical spots usually more strongly developed than in the male, and extensive basal amber may occur, sometimes extending as far as the nodus. The basal amber tends to become paler and eventually disappear with age, but the whole wing remains more or less enfumed.

The abdomen is marked with a lateral black stripe on either side, and two dorsal stripes. The latter grow longer with age, beginning in tenerals as two black dots at the base of each segment. In old females, they are triangular, with a broad base. Styli yellow with black apex.

Male grasping area: the hind rim of the pronotum is a plate, slightly depressed medially, with uplifted edges. The lam. mes. is triangular, with an anterior and lateral ridge, and smooth posterior border. The anterior carinal frame is a deep V that

Fig. 35: *Epallage fatime* (Charpentier, 1840); female
a. average extent of apical wing spot;
b. teneral female with typical pigmentation along main veins in wing basis

broadly widens towards the laminae. At the point of deflexion of the V a blunt tubercle occurs on either side of the frame.

Measurements (mm): *Male.* Total length 42–50; abdomen 28–37. *Female.* Total length 40–47 mm; abdomen 28–37 mm

Distribution: Macedonia, Greece, Thracia (Turkey-in-Europe), Anatolia, northern Iraq, Iran, and reaching Baluchistan in the east. In the south, it is found in Syria, the Lebanon, Jordan, and Israel. It occurs in most of the habitats where *Calopteryx* can also be expected, and is thus distinctly restricted to running water. Capture data range from March to September.

Israel (Locality records): Benot Ya'aqov bridge on Jordan River (1), Gadot (1), Dan (1), Tel el Qadi (1), Ḥula (1), Kefar Ḥittim (2), Wadi Bira (2), Buteicha Swamps (7), Deganya (7), 'Ubeidiya (7), Bitanya (7), Tabigha (7), Wadi Hammam (7), Ramallah (11), Wadi Fari'a (12), 'Ein Duyuk (13), 'En Gedi (13), "Dead Sea" (13), 'Ein es Sultan (13), Jericho (13), Wadi Qilt (13).

Also recorded from Ghor es Safi, from Wadi Mujib, Wadi Yabis, Wadi ez Ziqlab, and from Wadi Yarmouk near Kaziyé (Jordan-Syrian border).

Family LESTIDAE

Damselflies of small to moderate size with metallic or non-metallic body colours. The species with metallic colours perch with the wings half opened; those without metallic colours perch with the wings closed across the dorsum of the abdomen. Wings hyaline, no coloured spots, petiolated. Petiole ends at Ac; Ac meets Ab at hind border of wing. Only two an. d elongate and narrow, ending in a sharp distal angle. Sectors of arc arising at midpoint of arc. An oblique vein (o) always found between R_3 and IR_3 about midway between N and Pt. Pt elongate, usually about twice as long as wide. Synthorax with Su_1 reduced to a short trait close to the wing insertion. Abdomen slender; app. sup. of males always forcipate, spined externally.

1. Body colours not green metallic; base colour pale brownish, with numerous dark brown markings. Blue pruinosity, if present, restricted to wing implants. Pronotum with strong central lobe on posterior margin. d extremely narrow, and narrower in forewing than in hind wing, acutely pointed. Pt in forewing closer to wing apex than in hind wing by at least its own length. Males: ham. ant. (HA) leaf-shaped, rounded; ham. post. (HP) yellow, blade-shaped. **Sympecma** Burmeister
–   Body colours green metallic, with some yellow on sutures and ventral parts, and occasionally blue pruinescence and black stripes on parts of the body. Pronotum never with a strong central lobe on its posterior margin. Wings with d less narrow and similar in shape in forewing and hind wing. Pt in both wings at about the same distance from wing apex. Males: ham. ant. (HA) leaf-shaped, pointed; ham. post. (HP) black, hollowed-out. **Lestes** Leach

Genus SYMPECMA Burmeister, 1839
Handbuch Entomologie, 2:823

Sympycna Charpentier, 1840:19.

Type Species: *Agrion fusca* Vander Linden, 1820.
Small damselflies without metallic colours. Base colour ochre, extensively marked with brown. Ac situated closer to an_1 than to an_2, markedly oblique. d unusually narrow, and dissimilar in forewing and hind wing. Pt elongate, nearer to wing apex in forewing than in hind wing. Hind margin of pronotum with a prominent middle lobe. Perching animals hold their wings closed across the dorsum of the abdomen. Distribution: Europe, N. Africa, temperate Asia.
One regional species.

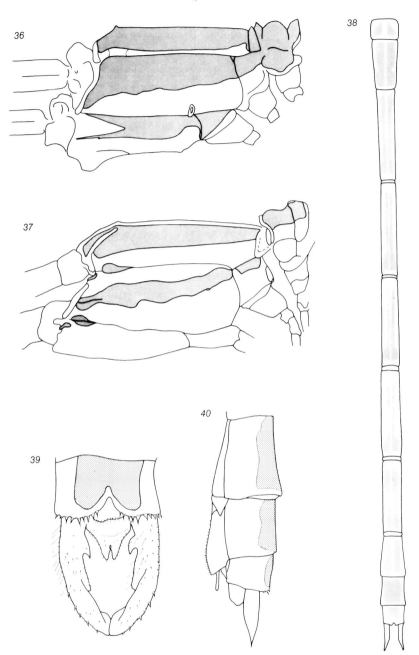

Figs. 36–40: *Sympecma fusca* (Vander Linden, 1820)
36–37. Maximum and minimum extent of dark bandings on synthorax
38. abdominal markings of a teneral male; 39. terminalia (anal appendages) of a male;
40. ovipositor of a female

Sympecma fusca (Vander Linden, 1820)

Figs. 36–40

Agrion fusca Vander Linden, 1820a:4.
Lestes fusca —. Selys & Hagen, 1854:161.
Sympycna fusca —. Selys, 1887:43.
Sympecma fusca —. Morton, 1924:30; Dumont, 1977b:134.

Male

Mouth parts, genae, labrum and clypeus yellowish. Frons heavily marked with black. Vertex and occiput black with narrow, sinuous, transverse yellow line at level of posterior ocelli. Rear of head yellow. Central part of occiput shallowly indented.

Pronotum largely black, lateral lobi yellow. Hind ridge with strong central lobe, bronze-black in colour, depressed; lateral lobi upright, yellow. Synthorax: ground colour light brown; carina yellow flanked by broad bronze bands. Humeral suture very narrowly black with apical widening and black patch on inf_2. A second black band with wavy edges between Su_1 and Su_2. Su_2 again very narrowly black, widening near its apical ending. Legs yellow with black spines and narrow external black stripe, sometimes interrupted and then forming a series of black spots.

Wings as for genus. Pt elongate, brown.

Abdomen: ground colour light brown, with bronze markings that widen near the top of each segment. In older specimens, most of the dorsum of S_{7-9} black. Appendages: app. sup. forcipate, with robust internal, subbasal tooth. Appendix narrowing slightly over half its length; 4–7 strong black spines externally. App. inf. closely apposed, conical, their tip surpassing the level of the tooth on the app. sup. Accessory genitalia: cleft in lam. ant. (LA) deep and narrow, its margins swollen; ham. ant. (HA) blade-shaped, yellow, rounded at tips, with a constriction in its external rim; ham. post. (HP) yellow, blade-shaped. Vesica spermalis elongate.

Female

Colour pattern similar to that in male. Pronotum: lam. mes. rather broadly triangular, the triangles deepened anteriorly. Carinal frame a wide V, bounded anteriorly by a straight plate with two posterior tubercles. Styli yellow, pointed, longer than S_{10}. Ovipositor yellow, rather short, not extending beyond half of S_{10}. Posterior third of v_3 with a row of blunt black spines on either side.

Measurements (mm): Male. Total length 34–38; abdomen 26–31. *Female.* Total length 34–38; abdomen 26–30.

Distribution: Palaearctic. Europe and western Asia. Species of the genus *Sympecma* hibernate as adults. They are therefore on the wing from the early days of spring till late autumn.

Israel (Locality records): A rare species, recorded from Nazareth (2), ♀, 23.V.1922 (Morton, 1924), and from Ḥadera (8; "Beth Gordon" Collection, Kibbutz Deganya A).

Genus LESTES Leach, 1815

Edinburgh Encyclopaedia, 9:137

Type Species: *Agrion sponsa* Hansemann, 1823.

Rather small damselflies with extensive green metallic markings on a yellow-ochre base colour; Ac situated about midway between the level of the two an. Pt always at least twice as long as wide; d similar in shape in both pairs of wings. Perching animals hold their wings half opened.

Distribution: Europe, Africa, N. America, temperate Asia. Four species are regional; two more are found in Turkey.

Key to the Species of Lestes
(Figs. 27, 41–48)

1. Rear of head yellow   2
   – Rear of head dark (except in freshly emerged specimens), usually green metallic or with blue pruinosity   3
2. Pt bicolorous, brown in its proximal half, clear in its apical half.
   Male: app. sup. yellow, with dark apices; app. inf. with divaricate apices.
   Female: valvifer of ovipositor posteriorly rounded or notched. Lam. mes. a narrow triangle, raised exteriorly into a crest.   **Lestes barbarus** (Fabricius)
   – Pt unicolorous, reddish-brown in fully coloured specimens (paler in freshly emerged ones), with narrow light yellow fringes.
   Male: app. sup. whitish; app. inf. short, slightly convergent, with rounded tips.
   Females: valvifer of ovipositor, laterally seen, produced into a backwardly directed point.
   Lam. mes. a narrow, fairly flat triangle.   **Lestes virens virens** (Charpentier)
3. Pt black or very dark brown, as long as 3–4 underlying cells.
   Male: app. sup. with broad inner expansion and one basal tooth.
   Female: lam. mes. and anterior rim of carinal frame raised.
                                                               Lestes macrostigma (Eversmann)
   – Pt brown, as long as 2 underlying cells.
   Male: app. sup. without broad inner expansion, and with two inner teeth.
   Female: lam. mes. and anterior rim of carinal frame not steeply raised.
                                                             Lestes viridis parvidens Artobolevski

Lestes barbarus (Fabricius, 1798)

Figs. 43, 47

Agrion barbara Fabricius, 1798:286.
Lestes barbara —. Selys, 1840:142.
Lestes barbarus —. Morton, 1924:30; Dumont, 1977b:135.

Type Locality: "Barbary", probably N. Morocco.

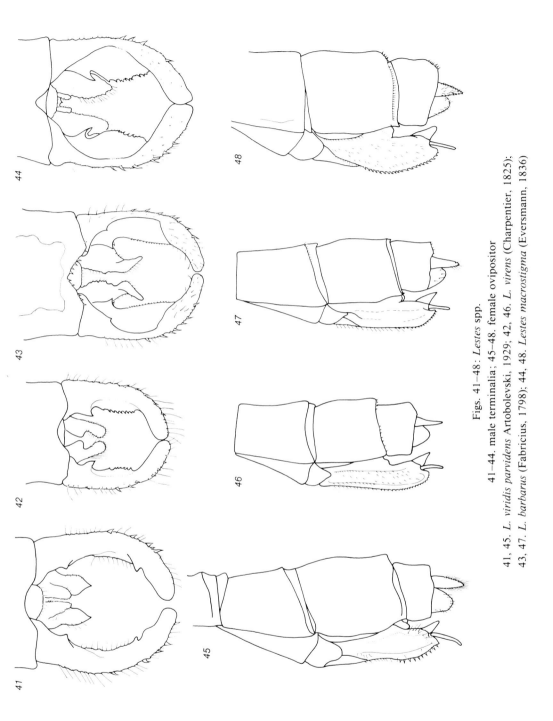

Figs. 41–48: *Lestes* spp.

41–44. male terminalia; 45–48. female ovipositor

41, 45. *L. viridis parvidens* Artobolevski, 1929; 42, 46. *L. virens* (Charpentier, 1825);
43, 47. *L. barbarus* (Fabricius, 1798); 44, 48. *Lestes macrostigma* (Eversmann, 1836)

Male

Dorsum of head bronze-green metallic; face and rear of head yellow. Yellow patches around the ocelli and the antennal implant.

Pronotum yellow marked with bronze. Thorax metallic bronze-green; carinal and humeral sutures yellow. epm_3 and $epst_3$ almost entirely yellow. Su_2 narrowly black. Legs yellow with black spines and dark tarsi. A black external stripe on all femora.

Wing venation light brown; Pt dark brown in its proximal half, light creamish in its distal half.

Abdomen: Dorsum of S_{1-8} marked with bronze, narrowly interrupted near the base of each segment. S_{10} with a central black spot, rest yellow. Sides of all segments yellow. Appendages yellow, tip of superiors dark. App. sup. with a basal spine, but no apical inner angle, and with a dorsal fold at the level of the inward curvature of the forceps. App. inf. less than half the length the superiors, apically pointed, divaricate, curved dorsad. Accessory genitalia as for genus.

Female

Colour markings as in male, but bronze-green replaced by bright metallic green. Abdomen: S_9 and especially S_{10} with contracted metallic markings on dorsum. Styli yellow. Ovipositor: v_3 with marginal denticles in its distal half. Valvifers quadrate, not distally invaginated at mid-ventrum.

Measurements (mm): *Male.* Total length 38–46; abdomen 28–35. *Female.* Total length 39–45; abdomen 26–33.

Distribution: Europe, where it is distinctly more common in Mediterranean climatic conditions than elsewhere. Extending far east into Asia, reaching India.

Israel (Locality records): Wadi Kurdana (4), Ramallah (11), Wasit (18), Birkat Bab el Hawa (18), Mt. Hermon (19). Capture data range from May till August.

Lestes virens virens (Charpentier, 1825)

Figs. 42, 46

Agrion virens Charpentier, 1825:8.
Lestes vestalis Rambur, 1842:250.
Lestes virens —. Morton, 1924:29; Schmidt, 1938:136.

Type Locality: Lusitania, i.e. Portugal and adjacent parts of Spain.

Male

Mouth parts, genae, postclypeus yellow (but a specimen from Hadera with postclypeus and labrum olivaceous). Dorsum of head dark metallic. Rear of head yellow, sometimes pruinose.

Pronotum green metallic, partly pruinose in mature specimens. Synthorax: dorsum metallic green, sometimes with cupric sheen. Humeral suture with or without yellow humeral stripe. In specimens from Hadera, this yellow stripe extended from the lam.

mes. to the alar sinus. In a long series from Birkat Bab el Hawa (Golan), this yellow band was narrowed by black stripes evolving on either side of the suture, eventually rupturing it near the alar sinus. In old specimens, the yellow stripe tends to become completely obliterated and replaced by a black stripe. $epst_2$ covered by metallic green, overshooting distally on epm_2. Margins of this stripe somewhat wavy. epm_3 yellow; a broad black stripe develops parallel to the one on the humeral suture. Very soon, however, it becomes covered by a layer of blue pruinescence. inf_2 yellow, with broad basal black spot. inf_3 yellow, later blue pruinescent. Metasternum yellow, with one or two pairs of ellipsoidal black spots. In senescent specimens, the metasternum turns very dark and pruinose. Legs yellow; spines and tarsi black, and a black stripe on the outer surface of the femora.

Wings: Venation light brown; Pt light to dark brown, its posterior and anterior rims bounded by a clear yellow vein. Costal and radial margins dark brown, almost black.

Abdomen green metallic with or without cupric sheen. End-rings of segments, and in old specimens their base, black. In younger specimens, base of S_{3-7} with yellow ring, medially interrupted by a black line. Appendages: app. sup. dark brown to black, with a basal tooth and moderately expanded, denticulated inner margin, not constricted (i.e. not forming an angle) at the level of the inward bend of the appendix. App. inf. yellow, short, apically rounded, and with convergent tips. Accessory genitalia: ham. ant. rather pointed posteriorly; ham. post. blade-shaped, their apex hollowed-out.

Female

Colour similar to that in the male, but cupric sheens more pronounced.

Abdomen S_{10} with styli yellow, shorter than segment. Ovipositor: v_3 denticulated. Valvifer produced posteriorly in lateral view, broadly embayed in ventral view.

Measurements (mm): *Male.* Total length 34–37; abdomen 26–28. *Female.* Total length 33–36; abdomen 25–27.

Distribution: Europe and South-West Asia.

Israel (Locality records): Found between April and July at Yir'on (1), Haifa (3), Migdal (7), Ga'ash (8), Ḥadera (8), Birkat Bab el Hawa (18).

Note 1: Subspeciation

On the Iberian Peninsula (*terra typica*) and in the Maghreb countries, *Lestes virens* always has a yellow humeral stripe, no black on epm_3, and reduced black on the metasternum. This represents the true *Agrion virens* of Charpentier. In temperate Western Europe, in Central Europe, and in the Balkans, a form is found in which the yellow humeral stripe is — at least — interrupted posteriorly, if not obliterated completely, while heavy black markings invade the sides and the ventrum of the synthorax. This form has been equated with *Agrion vestalis* Rambur, described from the environs of Paris (Schmidt, 1938), and is now widely considered as a subspecies of *Lestes virens*. However, Schmidt (*loc. cit.*) had seen specimens from Ḥadera (= Khedeira), all collected on the same day, in which the female was typically *virens*, but the male *vestalis*. Specimens seen by me from that same area agree very well with typical *virens*, but specimens from the Golan Heights displayed intermediate morphs as well as typical *virens* and typical *vestalis*. No subspecific status can thus be assigned to Levantine forms, but it is certainly worth placing on record that the local populations display a range of variability found only as extremes in the Western Mediterranean on the one hand, and in continental Europe on the other (see also hereafter).

Note 2: The status of *Lestes sellatus* Selys, 1862 (Selys, 1862:34)

A few males from "Syria and Egypt(?)", collected by Ehrenberg, and communicated to H.A. Hagen, were described by Selys (1862) as a separate species. The presumed type was re-examined by Schmidt (1938). It differs from *virens* in that the dorsum of the head is brownish-yellow (not metallic black). The app. inf. are said to be short, and each of them "almost forked", but presumed to be damaged. In the specimen seen by Schmidt, the app. inf. were lacking, but the interior dilated margin of the app. sup. narrowed apically in an angular fashion. The synthorax showed an unusually broad humeral suture, combined with extensive black markings on the sides. While the mention of "Egypt" is almost certainly misplaced, it remains possible that somewhere in the "Syria" of the nineteenth century, an as yet not rediscovered lestid occurs. On the other hand, it is quite possible that *L. sellatus* is nothing but a form of *L. virens*, combining the characters of the two described subspecies in an extreme form.

Lestes macrostigma (Eversmann, 1836)
Figs. 44, 48

Agrion macrostigma Eversmann, 1836:246.
Lestes macrostigma —. Selys & Hagen, 1850:150; Selys, 1862:296; Selys, 1887:42; Morton, 1924:30; Dumont, 1977b:135.

Type Locality: Orenburg, Russia.
A pruinescent species.

Male

Head: Mouth parts and genae yellow; rest of head, including rear side, black, except for some small yellow patches near the ocelli.

Pronotum black, covered by blue pruinosity. Legs black.

Wing venation black. Pt black, as long as 3–4 underlying cells.

Abdomen dark metallic (green, with bluish or cupric sheens). S_1 and S_2 in part pruinescent. Appendages: app. sup. forcipate, with basal spine, and a median internal expansion, set with fine spinules at its free border, constricted at the level of the curvature in the forceps. App. inf. less than half the length of the superiors, black, with broadly rounded apex, blade-shaped. Sides of the abdomen black, except for ventral margins of the tergites, and narrow end-rings of segments which are pale yellow. Rest dark green-blue metallic. Accessory genitalia as for genus.

54

Female

As male. Lam. mes. triangular, but raised steeply above the level of the synthorax. Anterior rim of carinal fork raised likewise. Ovipositor entirely black; valvifers not produced laterally, with a ventral, medially pointed invagination.

Measurements (mm): Male. Total length 40–47; abdomen 31–38. *Female.* Total length 39–45; abdomen 31–36.

Distribution: A species that favours brackish water for its larval development. In addition, it is thermophilic, and occurs in a fringe around the Mediterranean with only limited incursions into continental areas, except perhaps in the saline waters of the Pannonian plain. It is also found on major Mediterranean islands such as Cyprus and Sardinia.

Israel (Locality record): The only record available is from 'Atlit, S. of Haifa (4), in May (Morton, 1924).

Lestes viridis parvidens Artobolevski, 1929
Figs. 41, 45

Lestes viridis (Vander Linden). Morton, 1922:80; Morton 1924:30; Bartenef, 1925:56.
Lestes viridis parvidens Artobolevski, 1929:141. Schmidt, 1938:141; Dumont, 1972b:134.

Type Locality: The Crimea.
A non-pruinescent species.

Male

Head: Occiput, vertex and frons metallic green; yellow spots at the implantation of the antennae, around the ocelli, and on the hind rim of the occiput. Rear of head metallic. Anteclypeus black, sometimes with two yellow spots. Postclypeus and labrum greenish; labrum with dark fringe. Rest of mouth parts and genae yellow.

Pronotum yellow with median lobi dark metallic. Lateral lobi black. Synthorax: lam. mes. and anterior ridge of carinal fork largely yellow. Carina and humeral stripe yellow. Su_2 black. epm_3 and most of $epst_3$ yellow. Rest of synthorax bright metallic green. Legs yellow with black stripes along their entire length.

Wing venation brown; Pt relatively long, white in freshly emerged specimens, turning brown in mature animals.

Dorsum of abdomen green metallic. Sides yellow. End-rings of segments black. S_9 with a lateral, marginal and apical black stripe. S_{10} with an apical, squared embayment. App. sup. yellow with black tips. A basal tooth is present, but the inner margin is not dilated. The tip of the forceps is hollowed-out inwardly, and a small blunt tooth is implanted on the inner rim of the folding. Inferior appendages reaching to about half the length of the superior ones, black, conical, their apices upturned and pointed. Accessory genitalia as for genus.

Female
Colour pattern similar to that in the male.
Abdomen with styli black. The side of S_9 is yellow, surrounded by black; v_3 of the ovipositor also heavily marked with black. Teeth on ventral margin of S_3 comparatively strong. Valvifer rounded in lateral view; in ventral view, deeply hollowed-out.
Measurements (mm): *Male.* Total length 45–50; abdomen 34–39. *Female.* Total length 44–49; abdomen 35–39.
Distribution: The nominal subspecies inhabits west and central North Africa, and most of Europe, while ssp. *parvidens* is found in the Caucasian area, Anatolia, and in the Levant. Records range from April till October.
Israel (Locality records): Dan (1), H̱ula (1), Mt. Tabor (2), Na̱hal Tut (8).

Family PLATYCNEMIDIDAE

Damselflies of small size, coloured white or brown-red, marked with blue or black, sometimes pruinescent, not metallic. At rest, the wings are closed over the dorsum of the abdomen. Wings hyaline; Ac arising halfway between the two antenodals. Petiolation beginning at or proximal to Ac; Ab always present, continuing as A. d is a quadrilateral, elongate, widening distally. Abdomen cylindrical, narrow, sometimes slightly compressed in males. Superior appendages shorter than inferiors. Head transversely elongate, about twice as wide as synthorax.

Genus PLATYCNEMIS Burmeister, 1839
Handbuch Entomologie, 2:822

Type Species: *Libella pennipes* Pallas, 1771.
Small damselflies, with petiolated wings, a subquadrangular discoidal cell, and an oblique, diamond-shaped Pt. Pronotum with posterior lobe a broad plate in the male, and an isosceles triangle in the female. Lamina mesostigmalis triangular; anterior border of carinal ridge bulbously swollen in females. Males with app. sup. triangular; app. inf. longer and forcipate. Accessory genitalia: ligula with apical segment, flask-shaped, curved upwards, and with lateral flanges. Female with ovipositor relatively small.
Distribution: Europe, Asia, Africa.

Key to the Species of Platycnemis
(Figs. 12b–c, 49–55, 57–59)

1. Tibiae of 2nd and 3rd pairs of legs conspicuously dilated in both sexes. Male whitish or pale bluish with black markings; females brownish-reddish. App. sup. triangular with smooth apex. No pruinescence. **P. dealbata** Selys & Hagen
Tibiae not dilated. Teneral males whitish with black markings, very quickly turning dark blue to almost black due to the development of a layer of pruinescence. App. sup. triangular with bifid apex. Females coloured as males. **P. kervillei** (Martin)

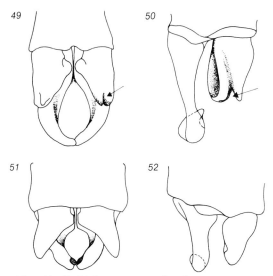

Figs. 49–52: Male appendages in *Platycnemis* spp.
49–50. *P. kervillei* (Martin, 1909); dorsal and lateral views
(arrow: subapical cleft in app. sup.);
51–52. *P. dealbata* Selys & Hagen, 1850; dorsal and lateral views

Platycnemis dealbata Selys & Hagen, 1850
Figs. 12c, 51–52, 54–55, 57–59

Agrion dealbata Klug, 1849:165 *nomen nudum* (in Selys & Hagen, 1850).
Platycnemis dealbata Selys & Hagen, 1850:165 (race of *acutipennis*), p. 388 (good species).
Platycnemis latipes race *dealbata* —. Selys, 1863:167 (formal description).
Platycnemis dealbata (race of *latipes*) —. Selys, 1887:46.
Platycnemis dealhata (sic!) —. Morton, 1924:30.
Platycnemis latipes dealbata —. Schmidt, 1938:141.
Platycnemis dealbata —. Dumont, 1977b:138.

Type Locality: "Egypt" (almost certainly misplaced!).

Note

Klug cannot have authority over the name *dealbata*, since he never published it himself. Information about the species was communicated by H.A. Hagen to Selys, and incorporated into *Revue des Odonates* (1850), which was published under the joint authorship of both dragonfly workers. On pp. 165–166 *dealbata* is considered, on evidence of a female, as a "race" of *P. acutipennis* (a Western Mediterranean species), but in a correction on p. 388, it is given full specific status. An important diagnostic character, the strongly dilated legs, is explicitly recorded here, and since no similar species is found in the Near East, we can be confident that what is currently referred to under the name *dealbata* indeed corresponds to this type material.

The type locality poses another problem. We do not know what is meant by "Egypt", but it is almost certain that *dealbata* does not occur in classical Egypt. Andres (1928) states that specimens from Sinai are present in the collection of the plant protection section of the Ministry of Agriculture, Cairo (collected by C.B. Williams), which might or might not be the area where the types were taken. It should be mentioned here that

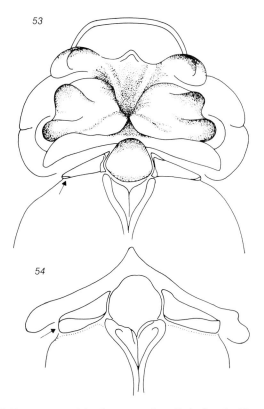

Figs. 53–54: Pronotum and lamina mesostigmalis in female *Platycnemis* spp.
53. *P. kervillei* (Martin, 1909); 54. *P. dealbata* Selys & Hagen, 1850
(arrows point to diagnostic differences on lamina mesostigmalis)

Hagen, already in 1849, had seen additional specimens from Syria, and had given them the manuscript name *P. syriaca* (see Selys & Hagen, 1850). On the other hand, early nineteenth century collectors often travelled to Egypt via the Levant, and many misplaced records have resulted from this. Among Odonata records from Egypt such species as *Calopteryx syriaca*, *Lestes viridis*, and *Coenagrion puella* fall into this category.

In conclusion, it appears that the *terra typica* of *P. dealbata* is not deltaic Egypt but either Sinai (from where no recent records are available), or the Eastern Mediterranean coastline.

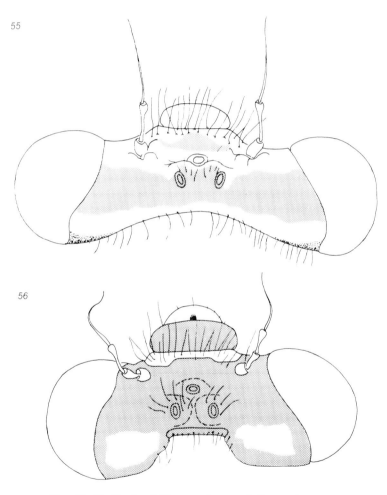

Figs. 55–56: Heads of *Platycnemis* sp. and *Coenagrion* sp.

55. *Platycnemis dealbata* Selys & Hagen, 1850;

56. *Coenagrion scitulum* (Rambur, 1842)

(with pear-shaped postocular spots)

59

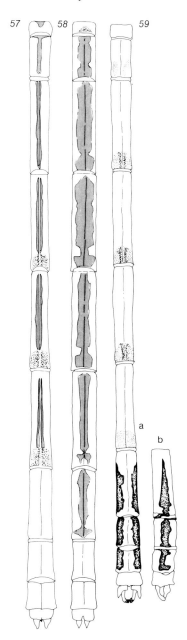

Figs. 57–59: *Platycnemis dealbata* Selys & Hagen, 1850;
abdominal markings
57. teneral female;
58. mature female;
59. a. mature male;
b. lateral view of terminal segments

Male

Mouth parts, labrum, genae greenish-white or with bluish sheen. Dorsum of head greenish white, with a black stripe on the frons, between the antennae, and a broad black transverse band on vertex and occiput. Rear of head pale. Antennae brown but pedicellum whitish-green.

Prothorax with pronounced relief, pale on the sides, but with extensive black markings, confluent to a large central patch on the middle of the pronotum. Hind rim of pronotum a wide plate, narrowing at its lateral edges. Synthorax: laminae triangular, with hind ridge slightly raised, especially outwardly. Carinal fork deep, anteriorly bounded by a thick rim, not bulbously swollen. Most of carina clear, flanked by a broad black band on either side. Humeral suture covered by a black stripe, extremely narrow and locally interrupted in teneral specimens, widening in older specimens, but rarely as wide as the clear antehumeral bands. A black stripe may also appear at the level of the meso-metathoracic suture, connected anteriorly to the humeral stripe. A very fine black line may, finally, appear on S_2 as animals age. Legs white or pale bluish with long black spines. A very fine black line on all femora, accentuated as a dark brown-black dot at the articulation of femur-tibia. Tibiae of second and third pairs of legs strongly dilated and flattened, pale, unmarked. Tarsal claws reddish-brown.

Wing venation brown. Pt light reddish-brown.

Abdomen pure white, or with a pale turquoise sheen. Segments 2–6 unmarked or with a diffuse darkening near the top of each segment. A fine mid-dorsal black line may be seen on S_2; S_{10} unmarked in young adults. S_7 first develops apical black markings, which later extend across S_8 and S_9 as bilateral black stripes. Eventually, a pair of lateral black dots may appear on S_{10}. Appendages: app. sup. triangular on a squared base. Apex rounded. App. inf. forcipate, with dark tips, considerably longer than superiors. Accessory genitalia: lam. ant. broadly excavated; ham. ant. broadly rounded posteriorly but pointed at inner apical corner; ham. post. small, inconspicuous. Vesica spermalis elongate, somewhat broadened posteriorly; membranaceous filling-aperture heart-shaped. Ligula with broad, blunt apex.

Female

Head entirely unmarked, reddish-brown, turning dark brown on maturity. Pronotum reddish-brown, developing black medial spots, in particular in the raised median sector of the hind ridge in old specimens. Shape of pronotum characteristic: the median lobes are divided into two tubercles by a median depression (the grasping area of the male's app. sup.). From the anterior collar, a V-shaped outgrowth projects backwards, deepened into a central pit. The hind ridge is medially produced into an upright, pointed tooth. Synthorax: lam. mes. triangular, the hind ridges of each triangle raised. Carina narrowly forked, with thick rims, deeply hollowed-out. Anterior rim bulbous, frame swollen. Synthorax cream-coloured, with brown markings, later stripes, developing in time. Abdomen, likewise, starting out wholly whitish-blue, turning cream brown, eventually two fine black stripes across the dorsum of S_{1-10} (Figs. 57–59). Legs with tibiae of 2nd and 3rd pair dilated, though

61

somewhat less than in males. Pure white in young specimens, they turn brown with age. In senescent specimens, a fine black stripe appears on the back side of the femora. Ovipositor relatively small; v_3 without spines; valvifer small. Pt white in young specimens, soon changing to a creamy colour, and to reddish-brown in fully mature specimens.

Measurements (mm): *Male.* Total length 28–37; abdomen 21–29. *Female.* Total length 30–38; abdomen 24–20.

Distribution: Central and eastern Anatolia, the Caucasus, Iran, Iraq, Afghanistan, northern India, Syria, the Levant.

It is one of the most widespread damselflies of Israel, and it is most commonly found between March and October, on virtually all running waters. In the south-west, it has been reported from Sinai in September (Andres, 1928), but precise localities are not known, and no specimens have been collected here in recent years.

Israel (Locality records): HaGosherim (1), Shamir (1), 'En T'eo (1), Sedé Nehemya (1), 'Ein Jalabina (1), Qiryat Shemona (1), Montfort (1), Nahal 'Ammud (2), Nahal Daliyya at Bat Shelomo (3), Ma'agan Mikha'él (4), 'Akko (4), Nahalal (5), Ashdot Ya'aqov (7), Migdal (7), Nahal Zalmon (7), Umm Juni (7), Kinneret (7), Hammat Gadér (7), Bitanya (7), Massada (7), Deganya (7), 'Ubeidiya (7), Tabigha (7), Wadi Samak (Nahal Samekh) (7), Binyamina (8), Rosh Ha'Ayin (8), Nahal Poleg (8), Ashqelon (9), Nabi Rubin (9), Jerusalem (11), Wadi Fari'a (12), Wadi Qilt (13), Qusbiya (18).

Also recorded from Nahr ez Zerka, and further from Wadi Wala, Wadi Hasa, Wadi Mujib, and all eastern tributaries of the Jordan River in Jordan. For Syrian records, see Al Hariri (1968) and Schneider (1981a); for Anatolian records, see Dumont (1977b).

Platycnemis kervillei (Martin, 1909)

Figs. 12b, 49–50, 53

Psilocnemis kervillei Martin, 1909:214. Gadeau de Kerville, 1926:79.
Copera kervillei —. Morton, 1924:30.
Platycnemis pennipes kervillei —. Schmidt, 1950a:82.
Platycnemis kervillei —. Schmidt, 1954a:66; Asahina, 1973:20; Dumont, 1977b:137.

Type Locality: Lake Homs, Syria.

Male

Head: Genae and mouth parts yellow. Labrum black in adults. Dorsum of head conspicuously black, covered by bluish pruinosity. In tenerals, head brownish, a brown-black stripe across the frons and a second, wider one, across vertex and occiput. Rear of head brownish.

Pronotum as in *P. dealbata*, yellow, very heavily marked with black in tenerals. Very soon, the entire prothorax turns uniformly black. Synthorax in teneral male marked with carinal, antehumeral, humeral, and Su_2 black stripes as in male *P. dealbata*. Again, the dorsum of the synthorax, down to the level of the suture between meso- and metathorax, soon turns uniformly black, covered by blue pruinescence. Metathoracic pleurites yellow, with black stripe over Su_2. Pt creamy in tenerals, bright brown in mature specimens. Legs pale yellow in tenerals, heavily striped along their entire length in adults. Tibiae not dilated.

Abdomen whitish with end-rings narrowly black and two small cuneiform oblique stripes at the base of each segment. S_{6-10} soon covered by black, and the entire abdomen becoming invaded by blue pruinosity in adults. Appendages entirely black, shaped as in *dealbata*, but superiors apically bifid. Genitalia as in *dealbata*, but filling aperture of vesica rounded posteriorly, not pointed.

Female

Light brownish when freshly emerged, with dorsum of head totally unmarked, and synthorax with carinal black stripe, humeral stripes, a faint black suture on Su_2, and two incomplete brown striae between Su_1 and Su_2. Legs yellow, with a series of brown spots along the femora. Abdomen uniformly pale. All these colours very soon turn black and pruinose blue, like in the male. Ovipositor as in *dealbata*. Styli short, shorter than S_{10}.

Pronotum: Hind rim more strongly raised than in *dealbata*, its lateral angles wider; lamina and carinal fork as in *dealbata*, but hind rim of laminar triangle not raised into a crest.

Measurements (mm): *Male.* Total length 30–35; abdomen 24–27. *Female.* Total length 30–34; abdomen 25–28.

Distribution: East Anatolia, northern Iraq, Syria, the Lebanon. The species has not been found in the Jordan Valley, but it occurs in the southern Litani Valley and in numerous localities in Syria. It thus typically represents a species bounded in the south by the Nehring line. *P. kervillei* is a late spring species that has been found between April and late July. It occurs on stagnant or slow-running waters. In habitus, it comes closest to *Pseudagrion syriacum*, while structurally it is very near to the western Anatolian and European *Platycnemis pennipes* (Pallas).

Family COENAGRIONIDAE

Rather small damselflies. Thorax with Su_1 reduced, visible only in its upper part. Wings hyaline, petiolated, with wing venation fairly open and most cells quadrangular. Two an. Ac arises halfway between the an, meeting Ab at the wing border (not in regional species) or some distance from it. d always untraversed, with lower distal angle acute, and costal side much shorter than cubital side. Tibiae of legs never dilated. Male accessory genitalia with ham. ant. large, squared, pointed at its inner anterior angle; ham. post. small, inconspicuous.

Key to the Genera of Coenagrionidae
(Figs. 11a, 13, 56, 60–203)

1.   arc situated distal to level of an_2. Very small species.   **Agriocnemis** Selys
   -    arc situated at level of an_2 2
2.   Dorsum of head entirely black, without postocular spots, or postocular spots present in teneral specimens only.   **Erythromma** Charpentier
   -    Dorsum of head black with blue or green postocular spots, or dorsum of head so sparsely marked with black that no such spots are defined   3
3.   Frons with an angular crest. Ground colour of abdomen bright red, with or without markings in the female. Some female forms largely or entirely black, but without vulvar spine.   **Ceriagrion** Selys
   -    Frons gently rounded, not angular. Ground colour of thorax and abdomen blue, green, brown or orange, but never bright red. In females, a vulvar spine may be present   4
4.   Males   5
   -    Females   8
5.   Pt in forewing bicolorous, usually divided into a black and a clear part. Pt in hind wing unicolorous, clear.   **Ischnura** Charpentier
   -    Pt unicolorous and of same colour in both pairs of wings   6
6.   Ground colour of thorax and abdomen brown, green, or yellow with black markings. Azure blue, if present, only on S_{8-9} of abdomen. Eyes and dorsum of head brick red in some species. Body occasionally pruinose.   **Pseudagrion** Selys
   -    Ground colour of thorax and abdomen azure blue, marked with black. Eyes never red, body never pruinose   7
7.   App. inf. simple, not differentiated into an inner and outer branch, triangular in side view, with upturned apex.   **Enallagma** Charpentier
   -    App. inf. not triangular in side view, of compound structure, with a well-differentiated inner and outer branch.   **Coenagrion** Kirby
8.   A vulvar spine on the ventrum of S_8 9
   -    No vulvar spine on the ventrum of S_8 10
9.   Robust species. R_3 springs more than 4 cells distal to N in forewing, more than 3 cells in hind wing. Hind margin of pronotum an oblique, smoothly rounded plate, without medial differentiation.   **Enallagma*** Charpentier
   -    Slenderly built species. R_3 springs 4 cells distal to N in forewing, 3 cells distal to N in hind wing. Hind margin of pronotum with a well-differentiated middle lobe (except *I. senegalensis*).   **Ischnura** Charpentier
10.   Hind margin of pronotum sinuous, always with a more or less uplifted and protruding middle lobe, but without stylettes and epaulettes. Ground colour green or blue.   **Coenagrion** Kirby
    -    Hind margin of pronotum upright, straight, with stylettes projecting anteriad (sometimes reduced to a tubercle) and epaulettes. Ground colour of synthorax green, brown or reddish.   **Pseudagrion** Selys

---

* Females of *Ischnura senegalensis* may key out here. Refer to Fig. 146. In *I. senegalensis* the vulvar spine is much more strongly developed than in *Enallagma cyathigerum.*

Type Species: *Agriocnemis lacteola* Selys, 1877.

Extremely small and fragile damselflies, ranking among the smallest species known. Colours non-metallic, usually blue or green marked with black, occasionally bright red. Females often occurring in several colour forms. Wings hyaline. Pt small, covering less than one cell. Usually 5–6 pn, rarely 8–9. d acutely pointed at distal lower end. arc situated distal to level of an_2. Ab arises proximal to Ac; Ac thus not springing from rim of wing. Head narrow, frons rounded. Postocular spots present. Posterior lobe of pronotum often with conspicuous differentiations. Abdomen slim, somewhat dilated terminally. Legs short. Female without vulvar spine.

Distribution: Asia, Oceania, Australia and Africa. Confined to tropical and subtropical climates.

One regional species.

Agriocnemis sania Nielsen, 1959

Figs. 60–71

Agriocnemis pygmaea —. Morton, 1924:34.
Agriocnemis sania Nielsen, 1959:33. Dumont, 1974:125.
Agriocnemis pygmaea sania —. Pinhey, 1974:225.

Type Locality: Oasis of Ghat, Fezzan, Libya.

Male

Mouth parts and anteclypeus pale yellow. Frons black, usually with two minute lateral green dots. Labrum black, metallic. Postclypeus, vertex, occiput black. A narrow yellow margin fringes the occiput. Postocular spots present, rounded, green.

Pronotum black with metallic purple sheen, its sides and hind rim yellow. Hind rim trilobate, with small side-lobes, and strong, upright, yellow middle lobe, deeply hollowed-out anteriorly. Synthorax black with purple sheen. Antehumeral stripes narrow, green. epm_2 black, except for some green on anterior half and on inf_2; epm_3 and $epst_3$ greenish yellow. Legs greenish-yellow, exterior surface somewhat darkened. Spines black.

Dorsum of abdominal segments 1–6 largely black. S_{1-3} greenish laterally, changing to brick red on subsequent segments. S_7 greenish-yellow laterally, basal half of dorsum black, apical part brick red. S_{8-10} red dorsally, green laterally. Appendages: app. sup. subquadrangular, with a downwardly pointed, foliate process, seen laterally. In posterior view, inner margin arcuate; foliate expansion appearing as a thin projection, outwardly directed. A strong basal spine on the lower inner corner, plus a tuft of

65

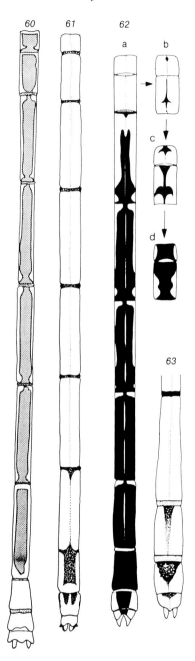

Figs. 60–63: *Agriocnemis sania* Nielsen, 1959; abdominal markings
60. male; 61. female, green form;
62. a. female, red form; b–d. female, red form,
evolution of markings on S_1–S_2 with time;
63. female, red form, abdominal segments 6–10 in aged specimens

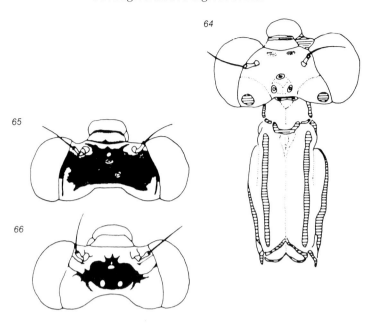

Figs. 64–66: *Agriocnemis sania* Nielsen, 1959
64. male colour pattern on head and thorax;
65. female, head, green form; 66. female, head, red form

golden hairs. Margins of S_{10} produced all around, but especially at the sides of the segment, hiding the short app. inf. App. inf. roughly triangular, ridged, with an oblique row of black spines near their base, a median fold, and a robust upper black spine.

Wing venation brown; Pt yellow; rest of venational characters as for genus.

Female

Two colour forms occur; the red one is common, the green one is rare.

Red form: Black markings very much reduced. Head entirely orange (but mouth parts yellow), except for a black patch around the ocelli. Pronotum orange. Some black on hind ridge only. Synthorax orange-red, sides changing to yellow. Abdomen orange; end-rings of segments and dorsum of S_{8-10} with some black. Legs: femora orange, rest yellow. Spines black.

Green form: darker. Clypeus and dorsum of head copiously marked with black. Abdomen entirely black dorsally, except for S_{8-10} where the black colour is reduced to spots. Sides of synthorax and of abdomen apple green.

Structure of pronotum and lam. mes. (Fig. 69): hind rim of pronotum trilobate, with lateral lobi smaller than central lobus, the latter flanked by two dorso-lateral tubercles of variable shape. Lamina mesostigmalis triangular, of complex relief. Carinal fork wide, each of its two arches widening into a triangular plate, bounded by an anteriorly raised rim, consisting of two adjacent tubercles.

Measurements (mm): *Male.* Total length 20–22; abdomen 16–17. *Female.* Total length 22–23; abdomen 17–18.5.

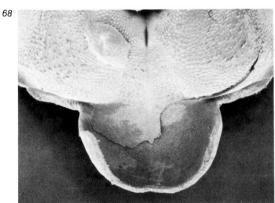

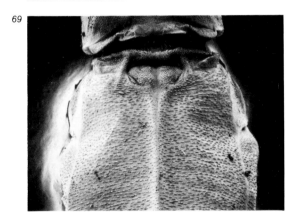

Figs. 67–69: *Agriocnemis sania* Nielsen, 1959
67. male terminalia, posterior view; 68. hind ridge of male pronotum;
69. female pronotum and lamina mesostigmalis, dorsal view

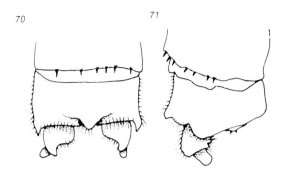

Figs. 70–71: *Agriocnemis sania* Nielsen, 1959; male terminalia,
dorsal and lateral views

Distribution: In the type locality, the species has been exterminated by the introduction of *Gambusia*. Pinhey (1974) records it from Ethiopia and from N. Kenya. There are no records from deltaic Egypt, but the species occurs in Sinai, and reaches the Jordan Valley.

Israel (Locality records): Benot Ya'aqov bridge on the Jordan River (1), Lake Ḥula (1), Mt. Tabor area (5), Petah Tiqwa (8), Hadera (8), Bet She'an (7), and Quseima Oasis (17).

Specimens have been caught between March and September.

Genus ISCHNURA Charpentier, 1840
Libellulinae eur., p. 20

Type Species: *Agrion elegans* Vander Linden, 1820.

Damselflies of small size, slender. Colours non-metallic, usually green or blue, sometimes brown, extensively marked with black. Females polychromatic. Wings hyaline. Pterostigma of males of different shape and colour in both pairs of wings, that of forewing bicolorous, short, covering one cell or less. d acutely pointed at its distal end. arc situated at an_2; Ab arising proximal to Ac. Head narrow; frons without angular crest. Postocular spots present. Posterior rim of pronotum mostly with prominent central lobe. Abdomen moderately short, segment 10 with distal rim raised to form one or two tubercles. Legs short. Female with a vulvar spine. Accessory genitalia: Lamina deeply cleft; ham. ant. squared, with inner anterior spine; lam. post. small. Ligula: the apical segment (glans) strongly built, often roughly triangular in lateral view, with long flanges and a pair of stylettes.

Distribution: Cosmopolitan, except for South America. Five species are regional.

Key to the Species of Ischnura
(Figs. 72–115)

1. Pt in forewing and hind wing differing in shape, that of the forewing being larger. Spine present on the inner-posterior corner of the lam. mes., adjacent to the carinal fork.

    Ischnura pumilio (Charpentier)

–   Pt in forewing and hind wing somewhat differing in shape, but not in size. No spine on the lam. mes., adjacent to the carinal fork   2

2. Males   3

–   Females   6

3. App. inf. much longer than app. sup. and about as long as S_{10}, forcipate   4

–   App. sup. and app. inf. of about equal length; these appendages not prominent, not forcipate, and much shorter than S_{10} 5

4. Hind ridge of pronotum with middle lobe modified into a steeply erected lamina; app. sup. crossed-over at their lower tips.   **Ischnura elegans ebneri** Schmidt

–   Hind ridge of pronotum with middle lobe inconspicuous, rounded, not erect. App. sup. with tips not crossed-over.   **Ischnura senegalensis** (Rambur)

5. Pt in forewing black in its middle, surrounded by a white fringe. App. sup. somewhat horse shoe-shaped, its inwardly and downwardly curved apex blunt.

    Ischnura evansi Morton

–   Pt in forewing black in its basal half, clear in its apical half. App. sup. with a finger-shaped expansion on its upper inner side; the fingers of the left and right appendix usually overcross each other.   **Ischnura fountainei** Morton

6. Vulvar spine strongly developed, long, produced over S_9. Hind rim of pronotum without any conspicuously developed middle lobes.   **Ischnura senegalensis** (Rambur)

–   Vulvar spine acutely pointed, but not produced over S_9. Hind rim of pronotum always with a well-differentiated, upright middle lobe   7

7. Middle lobe of hind rim of pronotum an upright lamina, longer than wide, not curved backwards.   **Ischnura elegans ebneri** Schmidt

–   Middle lobe of hind rim of pronotum shorter than wide, sometimes rounded, deeply bent over backwardly   8

8. Middle lobe about half as high as wide, arising from the upper rim of the hind border of the pronotum; the lower rim forms two folds at the sides of the lobe, not confluent under the lobe. Lam. mes. triangular, with hind ridge raised above the level of the synthorax along its entire length. Carina not forked, but preceded by a squared depression. Hind ridge of the depression not markedly raised.

    Ischnura evansi Morton

–   Middle lobe less than half as high as wide, arising from the upper rim of the hind border of the pronotum. The lower rim forms a fold at the sides of the lobes that is continuous under the lobe. Lam. mes. triangular, with hind ridge raised outwardly, but fading out towards the squared depression in front of the carina. Hind ridge of this depression markedly raised.   **Ischnura fountainei** Morton

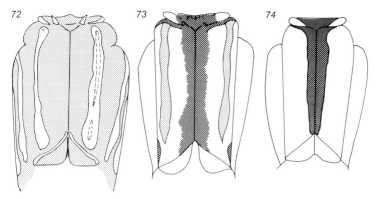

Figs. 72–74: *Ischnura pumilio* (Charpentier, 1825); synthoracic markings
72. male; 73. old female; 74. teneral female

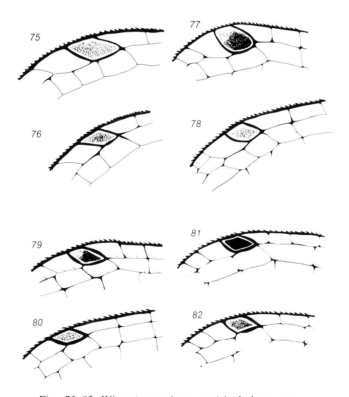

Figs. 75–82: Wings (pterostigma area) in *Ischnura* spp.
75–76. *I. pumilio* (Charpentier, 1825); female, forewing and hind wing;
77–78. *I. pumilio*, male, forewing and hind wing;
79–80. *I. elegans ebneri* Schmidt, 1938; male, forewing and hind wing;
81–82. *I. evansi* Morton, 1919; male, forewing and hind wing

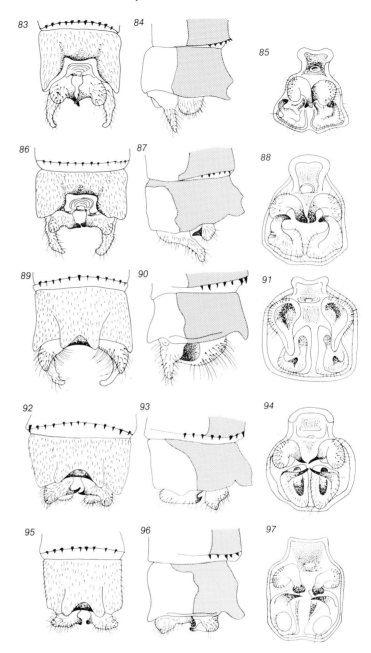

Figs. 83–97: *Ischnura* spp., male terminalia, dorsal, lateral, and posterior views
83–85. *I. elegans ebneri* Schmidt, 1938;
86–88. *I. senegalensis* (Rambur, 1842)
89–91. *I. pumilio* (Charpentier, 1825);
92–94. *I. fountainei* Morton, 1905;
95–97. *I. evansi* Morton, 1919

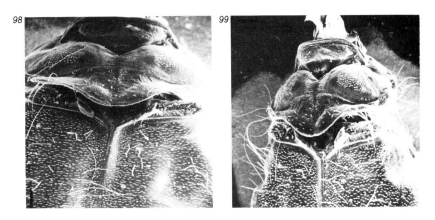

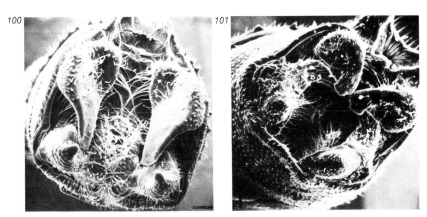

Figs. 98–101: *Ischnura* spp.
98. *I. pumilio* (Charpentier, 1825); female,
pronotum and lamina mesostigmalis; 99. the same, *I. pumilio*, male;
100. *I. pumilio*, male, terminalia, posterior view;
101. *I. evansi* Morton, 1919; male, terminalia, posterior view

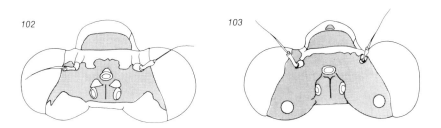

Figs. 102–103: *Ischnura senegalensis* (Rambur, 1842); head
102. teneral female; 103. aged female

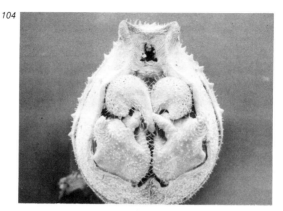

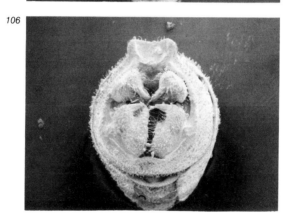

Figs. 104–106: *Ischnura* spp., male
terminalia, posterior view
104. *I. elegans ebneri* Schmidt, 1938;
105. *I. senegalensis* (Rambur, 1842);
106. *I. fountainei* Morton, 1905

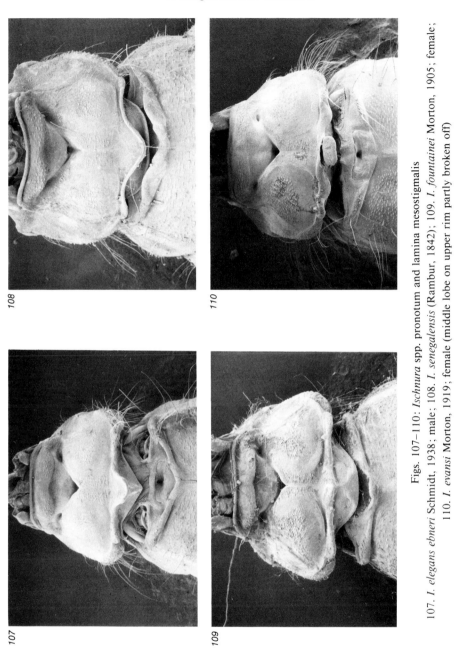

Figs. 107–110: *Ischnura* spp. pronotum and lamina mesostigmalis

107. *I. elegans ebneri* Schmidt, 1938; male; 108. *I. senegalensis* (Rambur, 1842); 109. *I. fountainei* Morton, 1905; female; 110. *I. evansi* Morton, 1919; female (middle lobe on upper rim partly broken off)

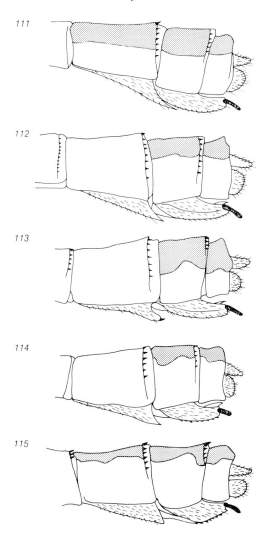

Figs. 111–115: *Ischnura* spp., female ovipositor
111. *I. senegalensis* (Rambur, 1842);
112. *I. pumilio* (Charpentier, 1825);
113. *I. elegans ebneri*, Schmidt, 1938;
114. *I. evansi* Morton, 1919;
115. *I. fountainei* Morton, 1919

Ischnura pumilio (Charpentier, 1825)

Figs. 72–78, 89–91, 98–100, 112

Agrion pumilio Charpentier, 1825:22. Selys, 1840:156.
Ischnura pumilio —. Selys, 1876a:267; Morton, 1924:32; Schmidt, 1938:142; Dumont, 1977b:140.

Type Locality: "Northern Italy".

Male
Mouth parts and clypeus yellowish. Dorsum of head black. Postocular spots large, blue or blue-green.
Pronotum black, its sides yellow or bluish. Posterior margin medially produced into a broad, rounded plate. Lam. mes. triangular, with a strong hook on the inner posterior corner. Carinal fork present between the laminae, with thick ridges. Synthorax black on dorsum, with a humeral stripe, and sides and ventrum bluish. Legs yellow, with dark external stripe.
Wings hyaline. Pterostigma in forewing black in its basal half, clear in its apical half. In the hind wing, it is unicolorous (pale yellow) and distinctly smaller than that in the forewing.
Abdomen black on dorsum, with end-rings narrowly yellow; posterior half of S_8 and most of S_9 blue, variously marked with black spots. Hind ridge of S_{10} only moderately raised and hollowed-out posteriorly. App. sup. shorter than app. inf., rounded in lateral view. In posterior view, they appear as two roughly triangular plates, rather acutely angled ventrally, somewhat concave medially. App. inf. longer than half of S_{10}, comparatively simple, forcipate.

Female
There are two basic colour forms.
Red form: Head bright orange, with isolated black spots on labrum and postclypeus. Vertex black. Postocular spots very large, orange-red, confluent across the dorsum of the head. Pronotum orange, with a large black patch at the base of the hind lobe. Free border of hind lobe orange. Legs orange, with interrupted black line on tibiae. Synthorax entirely orange. A carinal black line with wavy borders, and a very fine black stripe on the humeral suture. The partial suture between meso- and metathorax also accentuated by a fine black stripe. Abdomen S_1, S_2 and apical one-fourth of S_3 bright orange. Rest of abdomen black on dorsum, orange on sides. Styli short and robust, orange. As specimens age, the bright orange colours change to more brownish tinges; S_1 and S_2 become progressively covered by more extensive black markings, turning brown on their dorsum and greenish on the sides. The humeral space on the synthorax, likewise, turns chocolate brown.
Green form: The orange colour replaced by apple green. In old specimens, the green colours become darker; the postoculars shrink in size, their edges first turning deep brown. Eventually, they get separated from one another. The humeral space of the

synthorax becomes brownish, as well as the dorsum of S_1–S_2. The styli are brown. In both forms, the vulvar spine is strongly developed, and the ovipositor comparatively small. The Pt, light brown in both pairs of wings, is distinctly smaller in the hind wing than in the forewing. That in the forewing tends to be asymmetrical, in having the distal margin longer than the proximal one.

Measurements (mm): Male. Total length 30–33; abdomen 23–26. *Female.* Total length 25–34; abdomen 20–27.

Distribution: Europe, the Maghreb countries of North Africa, Asia Minor, West and Central Asia. The species is locally common in Turkey.

Locality records: Morton reports a male from 'Amman (30.VIII.1921) and another male from Wadi Samak [(Naḥal Samakh; 7), 13.V.1921]. Schmidt (1938) cites a male from Beharre, N. Lebanon, while Bolivar (1893) reports *I. pumilio* from "Syria". I have an additional record from Birkat Bab el Hawa, Golan Heights (18), June 1972.

Ischnura elegans ebneri Schmidt, 1938

Figs. 79, 80, 83–85, 104, 107, 113

Agrion elegans Vander Linden, 1823:104.
Ischnura elegans —. Morton, 1924:32; Fraser, 1933:351.
Ischnura elegans ebneri Schmidt, 1938:142. Schmidt, 1968:207; Dumont, 1977b:139; Schneider, 1981a:136.

Type Locality: Bethlehem (Solomon's Pool).

Male

Labium pale whitish, marked with black at base. Rest of mouth parts and genae largely greenish-blue. Anteclypeus blue, postclypeus steely black; dorsum of head black, with rounded blue postocular spots.

Pronotum black, sides of prothorax blue-green. Posterior rim with a robust middle lobe, steeply erect, longer than wide, black. Synthorax black to level of suture between meso- and metathorax; antehumeral stripes present, blue. Lower half of synthorax blue, blue-green or green. A black stripe over Su_2. Legs green or blue, with black stripes.

Wings hyaline; Pt of forewing diamond-shaped, its membrane white or pale blue, but its lower half strongly suffused with black. Pt in hind wing about the same size, its basal half not or only slightly blackened, and in any case not distinctly bicolorous and much paler than that in forewing.

Abdomen: Dorsum of S_{1-8} black. End-rings narrowly yellow. Sides of S_1, S_2 and sometimes S_3 blue. S_{3-7} with yellow sides. S_8 dorsally blue. S_{9-10} dorsally black, laterally blue. S_{10} with hind rim steeply raised, deeply cleft apically, thus forming the structure which apposes to the middle lobe of the female pronotum in copula. Appendages: app. sup. short, rounded in lateral view; in posterior view, with broad rounded body and

78

strong internal tooth. The inner teeth of both appendages almost invariably overcross each other (most diagnostic character of this subspecies). App. inf. much longer than app. sup. and more than half the length of S_{10}. Their base broadly triangular, with an inner blunt brown tooth, and external black, nipper-shaped projection.

Female

Much like the male, except for the genitalia and the Pt, which is pale yellow in both pairs of wings. Styli short, black. Ovipositor yellow. In young specimens, the sides of the synthorax are sometimes violaceous; in old specimens, the synthorax and the blue spot on the dorsum of S_8 may turn olivaceous or brownish. No heterochromatic form (without humeral black stripe, and with bright orange base colour of the synthorax) has been reported from the Levant (see further).

Measurements (mm): *Male.* Total length 29–36; abdomen 20–28. *Female.* Total length 30–37; abdomen 19–28.

Distribution: The subspecies occurs in south and central Anatolia, and in the Levant. Together with *P. dealbata*, it is the commonest damselfly of Israel, although the species does not reach beyond the Dead Sea. It has been found to co-occur with *I. evansi* on the Yarmouk River and in other localities in Syria and Jordan (Schneider, 1981a), and also in the oasis of El Azraq, Jordan (Dumont, unpublished observations), but not with any other of the regional *Ischnura*. It is on the wing from March to October, probably in several overlapping generations. In northern Anatolia and the Balkans, it is replaced by ssp. *pontica*.

Israel (Locality records): Dan (1), Yir'on (1), Sedé Neḥemya (1), 'Ein Jalabina (1), Qiryat Shemona (1), Ḥula (1), Hurshat Tal (1), Haifa (3), Zikhron Ya'aqov (3), Nahalal (5), Migdal (7), Naḥal Ẓalmon (7), Umm Juni (7), Lake Kinneret (7), Ḥammat Gadér (7), Bitanya (7), Massada (7), Deganya (7), 'Ubeidiya (7), Bet She'an (7), Ḥadera (8), Tel Aviv (8), Rosh Ha'Ayin (8), Bethlehem (11), Aqua Bella (11), Jericho (13), Ramat Magshimim (18), Wasit (18), Birkat Bab el Hawa (18).

Also very common on the wadis east of the Jordan River in Jordan, in the Lebanon and Syria (Schmidt, 1954a), and in Anatolia (Dumont, 1977b).

Ischnura fountainei Morton, 1905
Figs. 92–94, 106, 109, 115

Ischnura fountainei Morton, 1905:147. Ris, 1928:155; Kimmins, 1950:153; Asahina, 1973:18; Dumont, 1977b:140; Waterston, 1980:60; Schneider, 1981a:137.

Type Locality: Oasis of Biskra, Algeria.

Male

Mouth parts yellow. Labrum greenish with black base-line. Anteclypeus, genae and most of frons green. Postclypeus and dorsum of head black. Postocular spots small, blue.

Dorsum of pronotum black, sides yellow. Trochanters and femora broadly black, and a black stripe on all tibiae. Hind margin of pronotum with middle lobe almost completely reduced; a broad black lamella on the lower rim of the lobe forms a posteriorly directed plate with rounded margins. Synthorax: lam. mes. triangular. Carina not forked, but has a squared depression in front of it, between the laminae. Hind ridge of this depression markedly raised. Dorsum of synthorax black almost down to level of mesothorac suture. Antehumeral stripes narrow, interrupted, or almost completely obliterated. Su_2 black in its upper part. Metathorax olive green, with brownish darkenings on sutures.

Wings: Pt in forewing small; costal and radial veins much shorter than proximal and distal margins, thickened. Membrane whitish, but largely superimposed by black. In older specimens the whole Pt may turn black, with only a very fine whitish fringe. Pt in hind wing slightly larger, its four margins of roughly equal length, yellow with a very fine black reticulum, but not bicolorous.

Abdomen: S_{1-7} black dorsally; S_1, S_2, apical half of S_3 blue laterally; S_{3-6} yellow laterally and on top of each segment, where the black colour contracts. S_8 entirely blue; sides of S_{8-10} blue as well, their dorsum black. Hind margin of S_{10} only slightly raised, forked. Appendages very short, roughly of equal length. Superiors dark brown to black, inferiors yellow. Seen from behind, the app. sup. consist of a rounded outer tubercle, and an inner, slightly curved finger-like projection. App. inf. triangular, with a horizontally and inwardly directed black spine on top. Accessory genitalia as for the genus.

Female

Two basic forms occur. One is coloured more or less like the male (homochrome form), while the other has orange or brown as base colour, and lacks black humeral stripes (heterochrome form).

Homochrome form: this is by far the rarest. Coloured like the male: S_2 dorsally steel blue-black, and S_8 dorsally blue. Sides of synthorax greenish-brown. Pt yellow, parallelogram-shaped, identical in shape in both pairs of wings.

Heterochrome form: base colour of young specimens bright orange on the dorsum of the head, which has relatively small postocular spots. Legs entirely bright orange. Pronotum orange, with some black in the middle of the anterior collar, and at the base of the middle lobe of the hind margin. Synthorax bright orange, with a median humeral black stripe, divided into two halves by an orange carinal suture. Abdomen with S_{1-2} dorsally and laterally orange; a black marking on S_2. S_{3-7} black dorsally, yellow laterally. S_8 orange, but often considerably darkened by a mid-dorsal black marking. The same is true for S_9, the sides of S_{8-9}, and the whole of S_{10}. Styli and ovipositor orange. As specimens age, the postocular spots turn green, a series of black spots appear on the tibiae, ultimately confluent into a black stripe, the humeral area of the synthorax becomes a black stripe, and the humeral area of the synthorax turns chocolate brown, while the sides and dorsum of S_{1-2} become olive green. More brown appears on the terminal segments. The styli darken considerably and the ground colour of the abdominal segments changes to brown. No structural difference exists

80

between the two colour forms. The upper rim of the hind ridge of the pronotum shows a small median lobe, tongue-shaped, much less than half as long as wide. The lower rim of the ridge is developed into a triangular plate that is widest just beneath the middle lobe.

The lam. mes. are triangular and the posterior rim of each triangle is elevated outwardly, depressed medially. The carinal fork is ellipsoidal, deep, and its hind ridge is conspicuously raised above the level of the adjacent carina and lam. mes.

Measurements (mm): *Male.* Total length 29–34; abdomen 22–25. *Female.* Total length 27–34; abdomen 21–25.

Distribution: Northern Algeria, S. Tunisia, Libya, Egypt (oases of the Western Desert; no records from the Nile Valley), Saudi Arabia, Jordan, Iraq, Iran, and part of the Caspian area. Specimens have been captured between March and September.

This species frequently co-occurs with *I. evansi* and I. senegalensis (see below).

Israel & Sinai (Locality records): Jericho (13), 'Ein Fashkha (13), 'En Avedat (17), Quseima Oasis (21), Wadi Tala near Gebel Katharina (22), Et Tur (23).

East of the Jordan River, the species ranges further north, reaching the oasis of El Azraq and the Yarmouk River.

Ischnura evansi Morton, 1919
Figs. 81–82, 95–97, 101, 110, 114

Ischnura evansi Morton, 1919:146. Morton, 1924:32; Ris, 1928:155; Kimmins, 1950:153; Asahina, 1973:17; Waterston, 1980:59; Schneider, 1981a:136.

Type Locality: Basra and Amara, lower Iraq.

Male

Mouth parts and genae greenish, labrum green with broad basal black band. Postclypeus and frons green. Anteclypeus, vertex and occiput black. Small blue spots may be found in front of each ocellus. Postocular spots relatively small, blue. Hind ridge of head yellow.

Dorsum of pronotum black, with some minute yellow spots. Sides and margins of hind rim greenish-yellow. Median lobe well developed, upright, about half as long as wide. Adjacent to it, hind margin sinuously depressed, then raised again in a small lateral lobe. Lower rim of hind margin produced into a narrow plate, contiguous across the middle lobe, but at least slightly concave here. Lam. mes. broadly triangular; hind ridge of each triangle raised along its entire length. Hind ridge of carinal fork moderately raised. Synthorax with dorsum black. Black humeral sutures wide, but antehumeral stripe may be as wide as humeral stripe and is never obliterated. Sides of synthorax greenish or bluish. Some black on mesothoracic suture and on Su_2. Legs yellow, heavily marked with black on femora, and striped on tibiae.

Wings: Pt in forewing black, surrounded by a narrow yellow reticulum. Pt in hind wing yellow with very fine dark reticulum.

Abdomen: S_{1-7} black dorsally, S_{1-2} blue laterally, S_{3-7} yellow laterally, S_8 blue dorsally. Hind rim of S_{10} strongly raised, deeply bifid, the whole segment laterally elongate. App. sup. dark brown to black, app. inf. light brown, both pairs very short and appearing in side view as an upper blunt tubercle (sup.), and a lower blunt tubercle (inf.). In posterior view, the app. sup. are bulbous, with an internal, downwardly bent, blunt black tooth on each appendix, the tips of which are slightly divergent. App. inf. block-shaped, with a deep longitudinal furrow. From the upper external corner, a robust black spine projects inwardly. The lower inferior corner is swollen, and forms a blunt boss. Genitalia as for genus.

Female

The female is polychromatic, but there are distinct differences from all other species, including the preceding ones.

Homochrome form: head, pronotum (althought often with rounded yellow spots on middle lobes) and legs coloured as in male. Synthoracic black bandings as in male but ground colour of synthorax and sides of S_{1-2} greyish-blue. Abdomen coloured as in male; S_9 (partly), S_{10} and styli brown in young specimens, black in mature specimens.

Heterochrome form: differs from all other regional ones by the fact that true black humeral stripes develop here with age. Moreover, the ground colour of the synthorax in young specimens is not bright orange as in *I. fountainei* or in *I. senegalensis*, but has a distinct tinge of ochre, or sometimes dark yellow. Humeral bands begin to develop near inf_2 and under the humeral suture. As they lengthen, a second stripe appears more distally and above the suture. With time, both bands expand and finally merge into a single humeral stripe. This whole process is accompanied by a darkening of the humeral area, which turns chocolate brown. S_1 is mostly coloured as the synthorax, but S_2 is always largely black dorsally. S_8 is brown dorsally, not blue. The pronotum has two broad brown spots on its middle lobe, near the anterior collar, and two kidney-shaped brown patches behind the middle lobe of the hind rim in young specimens. With age, the kidney-shaped spots disappear, and the patches on the middle lobes decrease in size. The postocular spots are round, not acutely angled anteriorly. When these spots are confluent, the rear of the occiput always remains black behind them. Structurally, both colour forms are identical. Pronotum: hind ridge with well-expressed, tongue-shaped middle lobe; its sides depressed, somewhat sinuous. Side lobes small. Lower ridge of hind margin expanded on both sides in the depression between middle and side lobe, but these expansions not confluent under the middle lobe. Lam. mes. triangular, the hind ridge of each triangle raised along its entire length. Hind rim of carinal fork not raised above level of laminae.

Dimensions as for *I. fountainei*.

Distribution: Iraq, Iran, Syria, Jordan, Saudi Arabia, Sinai, Western Desert of Egypt. The westernmost record is from Siwa Oasis (Kimmins, 1950). In the latter locality and all over Sinai and the Negev, it is in frequent association with *I. fountainei* and *I. senegalensis*, often in great population densities. On the wadis east of the Dead Sea, and as far north as the Yarmouk River, and in the oasis of El Azraq, *I. evansi* is found in the company of *I. elegans ebneri*.

Israel & Sinai (Locality records): Jericho (13), 'En Gedi (13), Naḥal 'Arugot (13), 'Ein Fashkha (13), 'En Avedat (17), Quseima Oasis (17), Wadi Hibran (22), Ḥammam Musa (23).

Ischnura senegalensis (Rambur, 1842)

Figs. 86–88, 102–103, 105, 108, 111

Agrion senegalensis Rambur, 1842:276.
Ischnura senegalensis —. Selys, 1876c:273; Sowerby, 1917:10; Morton, 1924:32; Ris, 1928:159; Andres, 1928:25; Morton, 1929:60; Fraser, 1933:348; Dumont, 1977b:141.

Type Locality: "Senegal", West Africa.

Male

Mouth parts, genae, postclypeus, most of frons yellow or greenish. Labrum yellow with black base. Anteclypeus black, as are vertex and occiput. In young specimens, some green around the individual ocelli. Postocular spots round and blue.

Pronotum black. Anterior collar and sides green. Hind border of pronotum with wavy upper rim. Middle lobe expressed as a moderately developed, triangular plate. Inferior rim developed into a broad lower plate, with more or less strongly upturned margins. The two laminae and the carinal fork confluent to form a single post-pronotal triangle, sometimes with a small depression at the level of the carina. Synthorax green or blue, with narrow antehumeral bands. Legs: External side entirely black, flexor side greenish.

Wings: Pt in forewing a parallelogram with most of the membrane black and only the apical corner pale coloured. Pt in hind wing of same shape as in forewing, yellow with dark reticulum.

Abdomen: sides of S_{1-2} blue, dorsum black, contracted on top of S_2. Sides of S_{3-7} yellow, dorsum black. S_8 entirely blue; sides of S_{9-10} blue, dorsum black. Anal appendages: app. inf. longer than app. sup; app. sup. rounded in lateral view. Seen posteriorly, they are U-shaped, with inner rim longer than outer rim, somewhat spatulate, black. App. inf. with green base and black apex, forcipate, simple. Accessory genitalia as for genus.

Female

Again, a homochrome and a heterochrome form are found, the latter going through a continuous colour change with age.

Homochrome form: coloured like the male.

Heterochrome form: clear parts of head, pronotum, synthorax, abdominal segments 1–2, and legs light orange. Postocular spots confluent with each other and with the rear of the head. A central black patch on the pronotum and a carinal black band on the synthorax. Later, this colour changes to light brown, and further to chocolate brown or to dark olive green. In old specimens, the true postocular spots turn blue as in the homochrome form, while the posterior area of the head remains ochraceous or dark orange. The legs may develop dark brown, almost black stripes on femora and tibiae. The upper rim of the hind ridge of the pronotum (all forms) converges medially into a point. The hind ridge of the pronotum is underlain by a narrow, duplicature on its lower rim. The laminae and carinal fork combine to form a large triangle owing to supression of the medial rims of the laminal triangles. Pt as in hind wing of male. Vulvar spine strongly developed, long and produced, projecting over S_9.

Dimensions as for *I. elegans ebneri*.

Distribution: This species has a very wide range, covering the whole of Africa (except the Sahara), Iraq, Iran, India and even Japan. It is common in the deltaic area of Egypt, and in Sinai, and reaches the Negev. It is presumably perennial in localities where the climate permits.

Israel & Sinai (Locality records): Ga'ash (8), Rosh Ha'Ayin (8) (two isolated, very northern records of the species, the only ones within the domain of *I. elegans ebneri*), 'En Gedi (13), Jericho (13), Quseima Oasis (17), 'En Avedat (17), HaMakhtesh HaGadol (17), El 'Arish (20), 'Ein el Furtaga (22), Wadi Tayiba (23), Sarabit al Khadim (23), Et Tur (23), Feiran Oasis (23).

Genus COENAGRION Kirby, 1890
Cat. Odonata, p. 148

Type Species: *Libellula puella* Linnaeus, 1758.

Damselflies of small to moderate size, colours non-metallic. Male always with azure blue as ground colour, marked with black. Females with green or blue as ground colours. Wings hyaline. Pt similar in both pairs of wings, slightly longer than wide. 8–11 an in forewing. d acutely pointed distally. arc situated at an_2. Ab present, arising well proximal to Ac, which is situated halfway between an_1 and an_2, or nearer to an_1. Head narrow, no frontal ridge.

Postocular spots mostly well developed. Posterior lobe of pronotum variously modified. Legs not flattened. Anal appendages variable, occasionally of complex structure. Female without vulvar spine. Accessory genitalia: lam. ant. deeply cleft; ham. ant. as for family. Ligula with apical segment curved back over basal segment. Flanges present.

Distribution: Europe, N. America, Africa, most of Asia.

Key to the Species of Coenagrion
(Figs. 13, 56, 116–141)

1. Male: app. sup. as long as or longer than S_{10}, simple (not differentiated into an internal and external branch), forcipate.
    Female: a bulb-like, spiny swelling flanks the lam. mes.   **Coenagrion lindeni** (Selys)

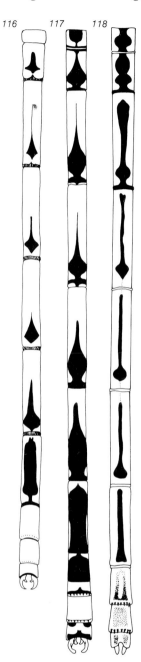

Figs. 116–118: *Coenagrion lindeni* Selys, 1840; abdomen
Fig. 116. dwarf male (Bet She'an);
117. normal male (Birkat Ram, Golan);
118. female (Birkat Ram, Golan)

–   Male: app. sup. shorter than S_{10}, and differentiated into an internal and external branch. Only the external branch is occasionally forcipate.

Female: no bulb-like swelling flanks the lam. mes.   2

2.   Male: app. sup. and app. inf. not markedly differing in length, although the former are longer than the latter.

Female: lam. mes. very broad, almost quadrangular, deeply hollowed-out.

Coenagrion scitulum (Rambur)

–   Male: app. sup. much shorter than app. inf.

Female: lam. mes. a narrow triangle, not markedly excavated   3

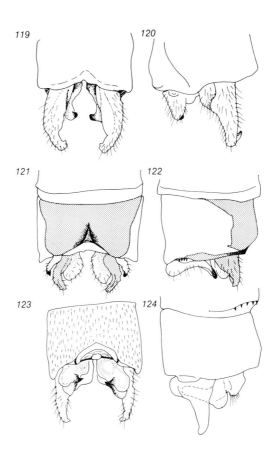

Figs. 119–124: *Coenagrion* spp., male terminalia,
dorsal and lateral views
119–120. *C. lindeni zernyi* (Schmidt, 1938);
121–122. *C. scitulum* (Rambur, 1842);
123–124. *C. puella syriaca* (Morton, 1924)

3. Male: tips of hook on apex of left and right app. sup. almost touching each other. Abdominal markings on S$_{3-6}$ with long central peak.

   Female: middle lobe of hind rim of pronotum notched.

   Coenagrion ornatum (Selys & Hagen)

– Male: tips of hook on apex of app. sup. widely separated. Abdominal black markings on S$_{3-6}$ with two lateral, but no central peaks.

   Female: middle lobe of hind rim of pronotum rounded.

   Coenagrion puella syriaca (Morton)

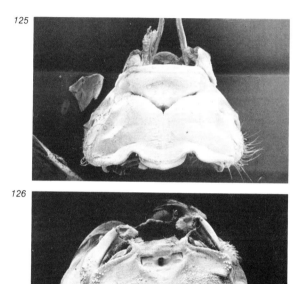

Figs. 125–126: *Coenagrion lindeni zernyi* (Schmidt, 1938); female
125. pronotum; 126. lamina mesostigmalis

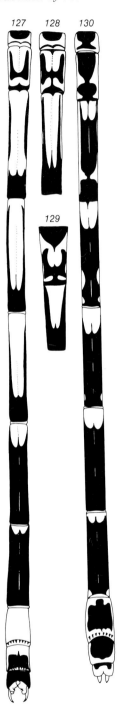

Figs. 127–130: *Coenagrion puella syriaca* (Morton, 1924); abdominal markings
127. male; 128–129. variation on S_1–S_3, male; 130. female

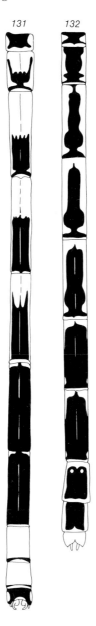

Figs. 131–132: *Coenagrion scitulum* (Rambur, 1842); abdomen
131. male; 132. female

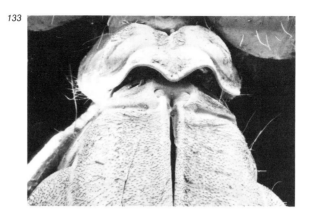

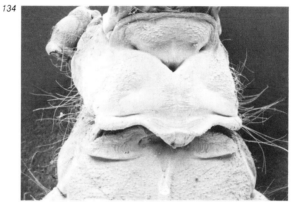

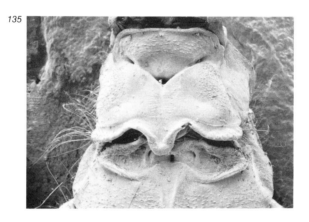

Figs. 133–135: *Coenagrion* spp., pronotum and lamina mesostigmalis
133. *C. puella syriaca* (Morton, 1924); female;
134. *C. scitulum* (Rambur, 1842); male;
135. *C. scitulum*, female

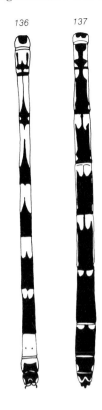

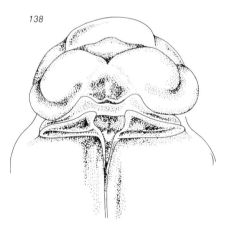

Figs. 136–138: *Coenagrion ornatum* (Selys & Hagen, 1850)
136. male abdomen; 137. female abdomen;
138. female, pronotum and lamina mesostigmalis

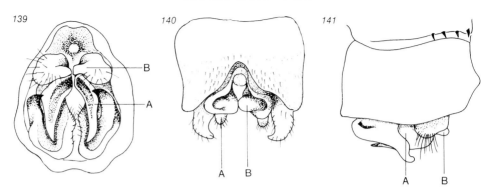

Figs. 139–141: *Coenagrion ornatum* (Selys & Hagen, 1850); male;
terminalia in posterior, dorsal, and lateral views
(A) blunt sclerotized tubercle on inner surface of app. sup.;
(B) ridge on downwardly directed tooth of app. sup.

Coenagrion lindeni (Selys, 1840)
Figs. 116–118

Agrion lindeni Selys, 1840:167. Morton, 1924:33.
Cercion lindeni —. Aguesse, 1968:107; Dumont, 1977b:141.

Two possible subspecies are regional. The males may be separated as follows:

1. Size large: abdomen 21–29 mm. Black stripes on meso-metathoracic suture and on Su_2. Black markings on S_2 confluent with end-ring of the segment, and black spear-shaped spots on S_{3-7} confluent with end-rings of each segment. S_8 dorsally blue with extensive black markings, or entirely black. **Coenagrion lindeni lindeni** (Selys)
– Size small: abdomen 19–23.5 mm. Black stripe on meso-metathoracic suture only present in upper part. No black on Su_2. Black spot on S_2 isolated from end-ring, and spear-shaped spots on S_{3-7} isolated from end-rings. S_8 dorsally blue.
Coenagrion lindeni zernyi (Schmidt)

Coenagrion lindeni lindeni (Selys, 1840)

Agrion lindeni Selys, 1840:167.
Coenagrion lindeni lindeni —. Schneider, 1981a:135.

Type Locality: "Belgium".

Male
Mouth parts, genae, labrum (except black virgule and base-line), postclypeus and frons blue. Anteclypeus black. Dorsum of head blue, with or without small blue spots

92

around ocelli, and elongate blue postocular spots, sometimes reduced to narrow lines, sometimes confluent across occiput, all in all rather variable.

Pronotum black, with blue spots on middle lobes, on sides, on lateral lobes of hind rim. Middle lobe broad, slightly produced. Laminae triangular. Carinal fork typical. Humeral black stripe narrower or at most as wide as blue antehumeral bands. Sides of synthorax blue, with fine black stripe on visible part of meso-metathoracic suture and on Su_2. Legs yellowish or bluish, with wide black stripe on back side of femora, and a narrow stripe on the tibiae.

Wings: Pt small (shorter than one adjacent cell), its distal side more oblique than its proximal side, yellow to light brown.

Abdomen blue, with black spear-shaped markings, confluent with end-rings of respective segments. Dorsum of S_8 black, its sides blue; S_9 entirely blue; S_{10} blue with black end-ring and a transverse black bar. Appendages: app. inf. less than half the length of app. sup., both pairs forcipate. Superiors long, longer than S_{10}, simple, somewhat broadened at about half their length.

Female

Head as in male, but ground colour green or ochraceous. Pronotum usually with more extensive clear markings than in male. Hind margin of pronotum wavy, with middle lobe somewhat depressed. Laminae broad, almost rectangular, with adjacent spiny hemispherical knob on pre-episternum. Antehumerals as wide as or wider than humerals. Sides of synthorax with black sutures as in the male on green, blue-green or blue ground colour. Dorsum of abdominal segments with broad sword-shaped markings on blue or brown ground colour. Dorsum of S_{9-10} blue, brown or darkened. Ovipositor rather robustly built; v_3 with a marginal row of spines. Wings as in male.

Measurements (mm): *Male.* Total length 30–36; abdomen 21–29. *Female.* Total length 30–36; abdomen 21–29.

Distribution: Europe, Anatolia, most of Syria.

The foregoing description applies to typical populations. I have included the nominal subspecies on evidence of its presence in Syria (Schneider, 1981a), and because of the occurrence on Birkat Ram (Golan Heights) of a peculiar form, which, although differing in some respects from topotypical material, stands closer to it than to *C. lindeni zernyi* (see below).

Coenagrion lindeni zernyi (Schmidt, 1938)

Figs. 116, 119–120, 125–126

Agrion lindeni. Morton, 1924:33.
Agrion lindeni zernyi Schmidt, 1938:143.

Type Locality: Ḥadera ("Khudeira"), Israel.

Male

Differs from the nominal subspecies by its smaller size (25–30 mm total length; 19–24 mm abdomen), by the reduced black on the lateral synthoracic sutures (no black on Su_2), by the abdominal black markings on S_{2-7}, not connected to the end-rings of the segments, and by S_{8-10} which are entirely blue. In some specimens, however, the antehumerals are narrower than in nominotypical specimens, and narrower than the humerals.

Female

Not described by Schmidt (1938). A single specimen examined by me from 'En Te'o differs from nominal examples by its size (total length 29 mm; abdomen 23 mm), by the antehumerals which are much narrower than the humerals, reduced black on the lateral sutures of the synthorax (a black stripe on the upper section of Su_2 only), and the dorsum of the abdomen, which is covered by broad black markings, including S_9 (dark brown) and most of S_{10} (brown). The specimen is homochrome.

Distribution: Limited to the upper reacher of the Jordan River and the coastal rivers of Israel, where it flies between March and October.

Israel (Locality records): 'En Te'o (1), Qiryat Shemona (1), Rosh Pinna (1), Lake Hula (1), Bet She'an (7), Deganya (7), Umm Junni (7), Bitanya (7), Bet Yerah (7), Hadera (8).

Status of the Golan Population of Birkat Ram (Figs. 117–118)

There is considerable variation in the extent of the black coloration on the abdomen, some specimens being close to *zernyi* in having the markings on S_2, S_3 and S_4 only narrowly or not at all connected with the end-rings of these respective segments. There is also variation in the extent of the black markings on the lateral thoracic sutures, but the antehumerals are wider than the humerals, and the specimens are robustly built, all well over 32 mm in length. The dorsum of S_8 is at least partly marked with black, but S_{10} is always blue. The females in this population were all heterochrome, with very narrow humeral stripes, narrow black abdominal markings, and the dorsum of S_{9-10} unmarked.

This population takes an intermediate position between the two subspecies. On evidence of size, the colours of the synthorax in the male, and the fact that S_8 is rarely entirely blue, I feel inclined to place it closer to the nominal subspecies than to ssp. *zernyi*, thus linking it up with typical populations that occur further north in Syria (Schneider, 1981a).

C. lindeni zernyi is a possible endemic of the Jordan Valley, although from the habitus of the Golan population, it appears that there is still genetical interchange with the nominal subspecies.

94

Coenagrion puella syriaca (Morton, 1924)
Figs. 123–124, 127–130, 133

Agrion puella syriaca Morton, 1924:32.
Agrion puella —. Gadeau de Kerville, 1926:78.
Agrion syriacum —. Schmidt, 1954b:236.
Agrion ponticum Bartenef, 1929:64.
Agrion pulchellum —. Gadeau de Kerville, 1926:78.
Coenagrion syriacum —. Dumont, 1977b:142.
Coenagrion puella syriaca —. Schneider, 1981a:135.

Type Locality: Zikhron Ya'aqov and Lake Ḥula, Israel.

Male
Mouth parts yellow. Labrum and anteclypeus black. Frons blue. Dorsum of head black. Postocular spots pear-shaped, sometimes confluent across hind ridge of head.
Pronotum black with blue spots on middle lobe. Hind margins laterally uplifted. Middle lobe depressed. Lamina a flat triangle, carinal fork typical. Synthorax with antehumeral blue bands narrower than black humerals. Side of synthorax blue with sutures marked in black, narrowly surrounded by yellow. Legs blue, heavily striped with black. Wings hyaline. Pt black, longer than wide.
Abdomen blue with black markings. S_2 with U-shaped spot, sometimes stalked and connected to end-ring of segment. Subsequent segments with basal black dots, from which two lateral peaks project anteriad. Appendages: app. sup. short, composed of two branches. The upper, inner lobe is short, black, and inwardly pointed. The lower, longer branch sends a strong hook backwards and between the app. inf. (visible in posterior view only). App. inf. as long as S_{10} and much longer than superiors. Their inferior internal angle short and rounded; upper external part produced into a long forceps. Accessory genitalia: lamina deeply but narrowly cleft; ham. ant. with long, pointed inner anterior corner. Vesica spermalis broad; filling area rounded posteriorly. Ligula with rather robust end segment and long curled flanges.
Female
A green and a blue form occur. Head, pronotum and synthorax with black markings as in male, but antehumeral stripes wider. Hind margin of pronotum sinuous, with centre depressed and sides uplifted. A rounded middle lobe is always defined. Laminae triangular, flat. Carinal fork very long and narrow. Abdomen with broad black markings on a green or blue background. Legs: femora less heavily marked with black than in male. Wings as in male.
Measurements (mm): *Male.* Total length 32 –37; abdomen 26–29. *Female.* Total length 30–36; abdomen 24–28.
Distribution: Israel, Syria, the Lebanon, Iran. In Anatolia, forms transitional to the nominal subspecies are found. The latter inhabits almost the whole of Europe. The intermediates seem to occur primarily in altitudinal biotopes, such as the Taurus

Mountains. In Israel, this is a spring species, limited to the upper Jordan Valley and coastal area.

Israel (Locality records): Ḥula (1), Gonén (1), Gadot (1), Naḥal 'Ammud (2), Zikhron Ya'aqov (3), Ḥadera (8), Aqua Bella (11).

Coenagrion scitulum (Rambur, 1842)
Figs. 56, 121–122, 131–132, 134–135

Agrion scitulum Rambur, 1842:266. Morton, 1924:33.
Coenagrion scitulum —. Dumont, 1977b:143.

Type Locality: Environs of Paris, France.

Male
Mouth parts yellow. Labrum blue with black base and median virgule. Postclypeus black. Frons blue. Dorsum of head black. Postocular spots large, elongate, blue, sometimes confluent.
Pronotum black. Anterior collar, part of sides, and part of hind rim blue. Posterior margin medially produced. Laminae triangular, broad, almost quadrangular, deeply excavated, their floor ribbed. Carinal fork small and shallow. Dorsum of synthorax heavily marked with black. Blue antehumerals always present, narrower than humeral black stripes. Lateral sutures also marked with black. Legs greenish-yellow, striped with black.
Wings: Pt elongate, yellow.
Abdomen blue, covered with black dots as in Fig. 131. Appendages: app. sup. somewhat longer than app. inf., of composite structure. Base bulbously swollen, with external hook-shaped projection. Inferiors much shorter; consisting of an oblique plate, with an upper and outer hook. Accessory genitalia as in *C. puella syriaca*.
Female
A blue and a green form occur. Head, pronotum, synthorax as in male, but antehumerals narrower and abdomen more copiously marked with black. Hind ridge of pronotum with middle lobe well expressed, produced backwards, sides of margin adjacent to it infolded. Lamina broad, quadrangular, deeply hollowed-out. Carinal fork very shallow, rather short.
Measurements (mm): *Male.* Total length 30–35; abdomen 24–27. *Female.* Total length 28–31; abdomen 21–24.
Distribution: Southern and Western Europe, and North Africa except Libya and Egypt. Eastwards extending to Iran. A late spring species, mostly found in June.
Israel (Locality records): Ḥula (1), Wadi 'Ayun (1), Newé Ativ (18).

96

Coenagrion ornatum (Selys & Hagen, 1850)

Figs. 136–141

Agrion ornatum Selys & Hagen, 1850:203. Gadeau de Kerville, 1926:78; Schmidt, 1954a:53.
Coenagrion ornatum —. Dumont, 1977b:143.

Type Locality: Hanover, W. Germany.

Male

Mouth parts yellow. A black stripe at the base of the labrum and at the base of the frons. Dorsum of head black. Postocular spots triangular, almost confluent across occiput, clear blue. Rear of head black, but a blue fringe on the back of the compound eyes.

Pronotum black, with anterior collar, two dots on median lobes, and fringe of hind margin blue. Synthorax: carina with broad black band. Blue antehumerals about as wide as black humerals. Sides of synthorax blue, with black stripes on upper (visible) sector of meso-metathoracic suture and on Su_2. Legs blue, with posterior black stripe. Wings: Pt black.

Abdomen blue. On S_2, a black marking in the form of a trident; on S_{3-6}, a black dot at the top of each segment, sending out a robust median peak anteriad. S_8 blue with two minute black dots. S_{9-10} largely black. Accessory genitalia as for genus. Appendages: app. sup. much shorter then app. inf., with a dorsal lobe that is apically hooked. Hooks of left and right appendage very close to one another. Lower lobe somewhat longer, rounded. App. inf. constricted at about half their length, ending in an inwardly curved hook.

Female

A blue and a green form occur. Built more robustly than the male, and with abdomen covered by extensive black markings over its whole length, contracting at the base of each segment only. Sides of abdomen blue, green, or yellowish-blue-green. Pronotum: hind rim with large lateral lobes, prominent and broadly rounded. Middle lobe much smaller, medially notched. The whole rim with a pale blue-green, almost white fringe. Adjacent lam. mes. narrowly triangular, flat.

Measurements (mm): *Male.* Total length 30–33; abdomen 23–26. *Female.* Total length 30–33; abdomen 24–27.

Distribution: From northern Iraq, across western Iran and Anatolia to Central Europe. Preferably flies over *Carex* marshes that have slowly running water. A spring species usually found in May and June. No records from the Jordan Valley are available, and it may not cross the Nehring line, but there is one record from marshes near Damascus, May 1908 (Gadeau de Kerville, 1926), that appears realistic.

Genus ENALLAGMA Charpentier, 1840

Libellulinae eur., p. 21

Type Species: *Agrion cyathigerum* Charpentier, 1840.

Small damselflies, with large postocular spots. Colours non-metallic. Males with terminalia usually short, with little or no differentia tions. Accessory genitalia: lamina deeply cleft; hamuli as in *Coenagrion*. Ligula with short flanges and no stylets. Females without stylets or epaulettes on the pronotum, but with a vulvar spine on the ventrum of S_8. Wings hyaline, arc close to an_2. Ab branches from Ac; Ac between an_1 and an_2.

Distribution: North America, Europe, Asia and Africa.

One species is regional.

Enallagma cyathigerum (Charpentier, 1840)

Figs. 142–149

Agrion cyathigerum Charpentier, 1840:163.
Enallagma cyathigerum —. Selys, 1876:496; Kirby, 1890:145; Selys, 1887:47; Gadeau de Kerville, 1926:78; Dumont, 1977b:141.

Type Locality: "Silesia".

Male

Mouth parts yellow. Labrum blue. Face blue but clypeus partly black. Dorsum of head black with blue postocular spots, round, triangular or cuneiform.

Pronotum black with blue fringes. Synthorax: a broad black carinal band, and black humeral stripes, widening anteriorly, but narrower than blue antehumeral bands. Sides of synthorax blue. Upper part of Su_2 black. Legs: flexor side blue-green, extensor side black.

Wings: venation as for genus. Pt oblique, rather large, dark brown.

Abdomen azure blue, with black markings. Marking on S_2 typically omega- or cup-shaped (whence the name of the species). S_{8-9} blue, S_{10} largely black. Accessory genitalia as for genus. Appendages: superiors much shorter than inferiors, with inner (deeper) lobe and outer (upper) lobe. Inferiors simple, triangular in side view, with apices pointed and upturned.

Female

A homochrome form co-exists with a heterochrome form with ground colour clear brown. Postocular spots and antehumerals usually wider than in the male. Pronotum with broad clear fringes, and two large blue or brown oval spots on the middle lobes. Hind margin an oblique, simply rounded plate. Abdomen broadly marked with black, this colour contracting at the base of each segment, especially on S_9, but not on S_{10}. Ovipostior rather small, but a robust vulvar spine present on the ventrum of S_8.

Measurements (mm): *Male.* Total length 29–36; abdomen 25–29. *Female.* Total length 30–35; abdomen 24–28.

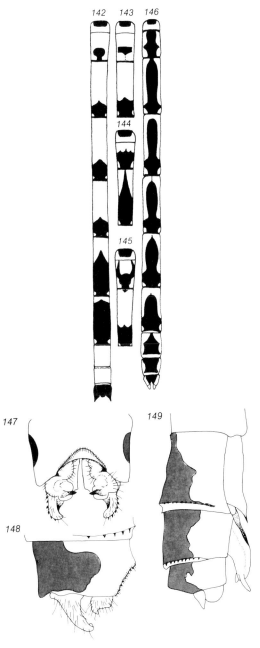

Figs. 142–149: *Enallagma cyathigerum* (Charpentier, 1840)
142. male, abdomen;
143–145. variation on S_{1-3}, male;
146. female, abdomen;
147–148. male, terminalia in dorsal and lateral views;
149. female, ovipositor

Distribution: A circumboreal species, widely distributed in Europe and Western Asia. Related species occur in North Africa and Central Asia. In the Caucasus, a possible subspecies (*E.c. rotundatum* Bartenef) occurs, but the species is typical on most of the Anatolian plateau (Dumont, 1977b). There is only one record from the Levant: marshes near Damascus in May 1908 (Gadeau de Kerville, 1926). The species might occur on the Golan Heights and Mount Hermon during summer.

Genus PSEUDAGRION Selys, 1876
Bull Acad. r. Belg., Ser. 2, 42:490

Type Species: *Agrion caffrum* Burmeister, 1839.

Medium-sized, rather firmly built damselflies, variously coloured, occasionally pruinose, non-metallic. Frons without transverse ridge. Postocular spots present, of variable size, but in some females head almost unmarked with black. Female pronotum mostly with a pair of stylets, sometimes reduced to very small tubercles. Lam. mes. and carinal fork variously modified. Sometimes a lateral epaulette present. No vulvar spine. S_{10} in males not raised into a terminal tubercle. App. sup. about as long as S_{10}, mostly differentiated into an inner and outer branch, the latter invariably ending in an inwardly bent hook. App. inf. usually shorter, spatulate or of complex structure. Wings. hyaline. Pt an elongate quadrilateral, similar in shape, size and colour in all wings. arc arises at an_2. Ac arises about halfway between an_1 and an_2 or closer to an_2. d with lower distal angle acute. Ligula: L_2 with a single pair of flanges, occasionally very short, never filiform.

Distribution: Afrotropical and Oriental regions.

Note

The regional representatives were reviewed in Dumont (1973). It was shown later, however, that *P. acaciae* Förster should be omitted and replaced by *P. niloticum* Dumont, an endemic of north east Africa (Dumont, 1978a). The latter species will probably not appear in Sinai and is not included in the present fauna.

Key to the Species of Pseudagrion
(Figs. 150–189)

1.	Males	2
–	Females	6
2.	App. sup. simple, no inner branch. Dorsum of head, pronotum and synthorax bright red. **Pseudagrion sublacteum mortoni** (Ris & Schmidt)	
–	App. sup. with inner and outer branch, separated by a ridge. Head and thorax variously coloured, occasionally pruinose, but never red	3
3.	Inner branch of app. sup. protruding well beyond the outer branch, finger-shaped. Dark insects, heavily pruinose, this pruinosity forming blue secondary antehumerals in mature specimens. Postocular spots small, rounded, green. **Pseudagrion syriacum** (Selys)	

100

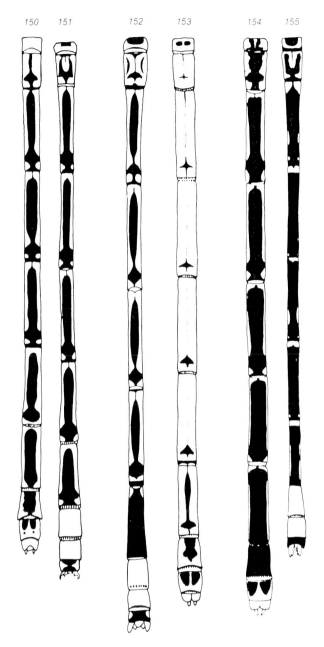

Figs. 150–155: *Pseudagrion* spp., abdominal markings
150. *P. torridum hulae* Dumont, 1973; female;
151. *P. torridum hulae*, male;
152. *P. sublacteum mortoni* (Ris & Schmidt, 1936); male;
153. *P. sublacteum mortoni*, female;
154. *P. syriacum* (Selys, 1887); female;
155. *P. nubicum* Selys, 1876; male

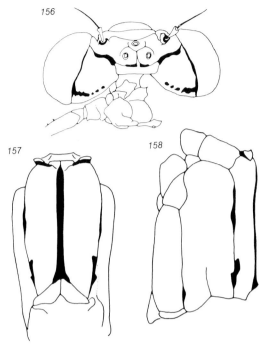

Figs. 156–158: *Pseudagrion sublacteum mortoni* (Ris & Schmidt, 1936); female
156. head; 157–158. synthorax, dorsal and lateral views

– Inner branch of app. sup. about as long as or shorter than outer branch. Green-and-black or blue-and-black species with reduced pruinosity. Postocular spots green or blue, fairly large   4

4. Inner and outer branches of app. sup. of about equal length. Inner margin of inner branch with a series of short teeth and a single, somewhat more strongly built sub-basal spine.
                                                                      Pseudagrion nubicum Selys

– Inner branch of app. sup. much shorter than outer branch, but strongly developed, with two strong thorn-shaped outgrowths, pointing medio-ventrad   5

5. Distal rim of inner branch of app. sup. smoothly rounded. Dorsum of head marked with black in such a way that postocular spot sare defined. Humeral and carinal black stripes relatively wide.   **Pseudagrion torridum torridum** Selys

– Distal rim of inner branch of app. sup. concave. Dorsum of head with black markings so reduced that the postocular spots are not separated from the other clear areas of the dorsum of the head. Humeral and carinal black stripes narrow.
                                                        Psuedagrion torridum hulae Dumont

6. Stylets reduced to a couple of very small tubercles on top of the hind margin of the pronotum.   **Pseudagrion sublacteum mortoni** (Ris & Schmidt)

– Stylets present as projections of the hind margin of the pronotum towards the median lobes of the pronotum   7

7   Stylets long, finger-shaped, reaching the median lobe of the pronotum. No epaulette 8

– Stylets short, apically pointed, not or slightly projecting beyond the hind lobe of the pronotum. A black lateral epaulette present.   **Pseudagrion syriacum** (Selys)

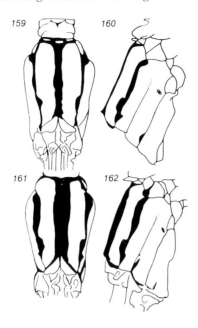

Figs. 159–162: *Pseudagrion* spp., synthorax, dorsal and lateral views
159–160. *P. torridum hulae* Dumont, 1973; male,
161–162. *P. t. torridum* Selys, 1876; male (Sinai)

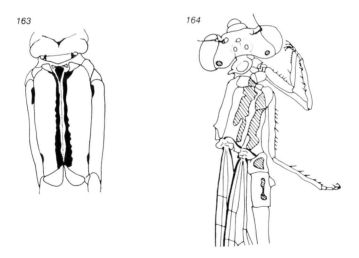

Figs. 163–164: *Pseudagrion syriacum* (Selys, 1887)
163. female, synthorax, dorsal view;
164. male, head, thorax and base of abdomen
(striped: clear areas)

Figs. 165–169: *Pseudagrion* spp., heads
165. *P. torridum torridum* Selys, 1876 (Sinai); male;
166. *P. torridum hulae* Dumont, 1973; male;
167. *P. t. hulae*, female;
168. *P. nubicum* Selys, 1876; male;
169. *P. syriacum* (Selys, 1887); female

8. Stylets extending more than halfway across middle lobe of pronotum. lam. mes. triangular, somewhat hollowed-out, its medial margin almost straight, not sinuously produced along carinal fork. Synthorax dark, antehumeral green stripe about half as wide as mesepisternum. **Pseudagrion nubicum** Selys
–   Stylets shorter than half the length of middle lobe of pronotum. Lam. mes. an irregular, hollowed-out, wide, ribbed triangle with medial margins sinuously produced against the greatly compressed carinal fork. Synthorax with more reduced black markings; width of antehumeral stripe exceeding half that of the mesepisternum   9
9. Carinal and humeral stripes relatively wide. Postocular spots usually well circumscribed by black. Egypt, south-western Sinai. **Pseudagrion torridum torridum** Selys
–   Carinal and humeral stripes narrow. Postocular spots not circumscribed by black. Lake Hula. **Pseudagrion torridum hulae** Dumont

104

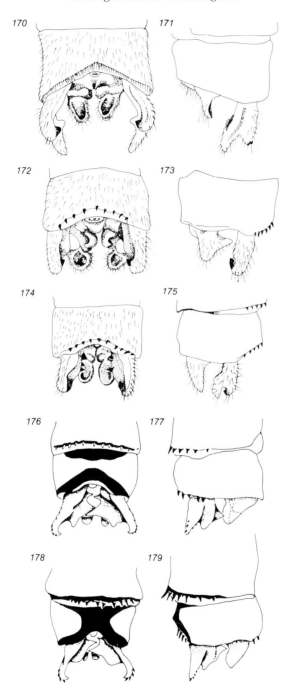

Figs. 170–179: *Pseudagrion* spp., male terminalia, dorsal and lateral views
170–171. *P. syraicum* (Selys, 1887); 172–173. *P. sublacteum mortoni* (Ris & Schmidt, 1936);
174–175. *P. nubicum* Selys, 1876; 176–177. *P. torridum torridum* Selys, 1876;
178–179. *P. torridum hulae* Dumont, 1973

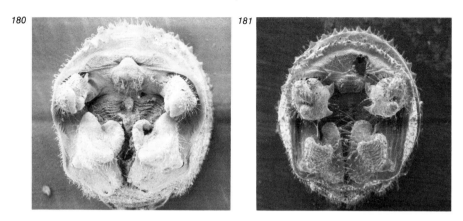

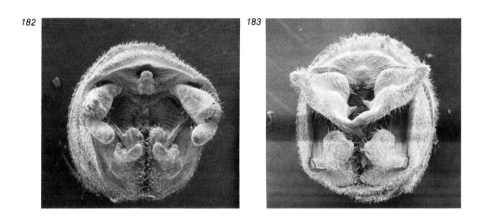

Figs. 180–183: *Pseudagrion* spp., male terminalia, posterior view
180. *P. sublacteum mortoni* (Ris & Schmidt, 1936);
181. *P. nubicum* Selys, 1876;
182. *P. syriacum* (Selys, 1887);
183. *P. torridum torridum* Selys, 1876

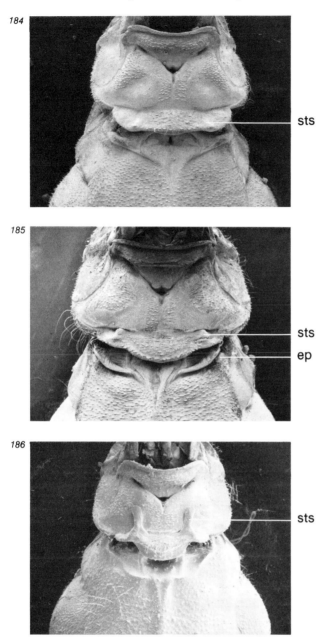

Figs. 184–186: *Pseudagrion* spp. females, pronotum and lamina mesostigmalis
184. *P. sublacteum mortoni* (Ris & Schmidt, 1936);
185. *P. syriacum* (Selys, 1887);
186. *P. torridum torridum* Selys, 1876
(sts=stylets; ep=epaulette)

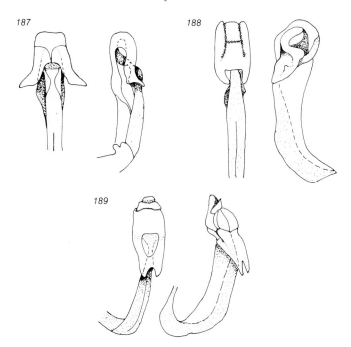

Figs. 187–189: *Pseudagrion*, spp. male, ligula
187. *P. syriacum* (Selys, 1887);
188. *P. sublacteum mortoni* (Ris & Schmidt, 1936);
189. *P. torridum hulae* Dumont, 1973

Pseudagrion sublacteum mortoni (Ris & Schmidt, 1936)

Figs. 152–153, 156–158, 172–173, 180, 184, 188

Agrion sublacteum Karsch, 1893:40.
Pseudagrion acaciae Förster, 1906a:56. Morton, 1924:34.
Pseudagrion Morton, 1929:63.
Pseudagrion mortoni Ris & Schmidt, 1936:55. Schmidt, 1938:146.
Pseudagrion sublacteum mortoni —. Dumont, 1973:175. Waterston, 1980:60.

Type Locality: Ghor es Safieh, Jordan.

Male

Labium, labrum dark orange; most of frons and dorsum of head bright red, including large confluent postocular spots. Synthorax red on dorsum, changing to green on sides. Carinal and humeral black stripes rather narrow. Ground colour of abdomen blue-green on S_{1-7}, marked with black as in Fig. 152, S_{8-9} blue, unmarked. S_{10} blue with black spot. App. sup. rather robustly built, compactly built, hooked apically, without an inner branch. App. inf. much shorter, roughly triangular in lateral views. In posterior view, a differentiation into an outer and inner half is observed. The outer part consists of two tubercles, separated by a groove, the lower one the larger; the inner part forms a less strongly developed ridge. Accessory genitalia: lamina with a deep V-shaped cleft. Hamuli as for genus. Ligula: L_2 with short broad flanges. Legs yellow, femora brown.

Wings: Pt yellow, elongate, distal margin more acute than proximal one.

Female

Face pale brown, with an orange tinge on the labrum. Black markings on head sparse. Thorax and abdomen largely pale brown. A greenish sheen visible on the sides of the synthorax in live specimens. Black stripes and dots on synthorax as in Fig. 157. Hind margin of pronotum almost straight, steeply raised. Stylets reduced to a pair of very small tubercles. Lam. mes. triangular, well raised above synthorax, with wavy posterior margin. Carinal fork wide, its ridges widened into two ellipsoidal plates. Abdomen sparsely marked with black. Legs yellow, variously striped with brown on femora.

Measurements (mm): Male. Total length 32–38; abdomen 26–31. *Female.* Total length 33–39; abdomen 27–31.

Distribution: The nominal subspecies is widespread in Africa south of the Sahara, while ssp. *mortoni* is restricted to the valleys of the Jordan River and its affluents in Israel, Jordan, and Syria, and to the short coastal rivers of Israel. However, recently Waterston (1980) has reported this subspecies from Saudi Arabia, a very remarkable range extension.

Israel (Locality records): Ḥula (1), Haifa (3), Kefar Ruppin (7), Yarmouk River (7), 'Ubeidiya (7), Naḥal Tanninim (8), Wadi 'Auja (13).

Also recorded from Nahr es Zerka, and from several localities in Jordan and Syria (Schneider, 1981a).

Pseudagrion syriacum (Selys, 1887)

Figs. 154, 163–164, 169–171, 182, 185, 187

Pseudagrion praetextatum var. (?)*syriacum* Selys, 1887:48.
Pseudagrion syriacum —. Kirby, 1890:153; Dumont, 1973:179; Schneider, 1981a:137.
Pseudagrion praetextatum Selys, 1876b:494. Morton, 1924:34; Gadeau de Kerville, 1926:78.
Pseudagrion kersteni (Gerstaecker, 1869). St. Quentin, 1965:538.
Pseudagrion kersteni (pars) —. Ris & Schmidt, 1936:18; Pinhey, 1964:25.

Type Locality: Beirut, Lebanon.

Male

Pale brown when freshly emerged, but quickly turning dark grey, almost black. Face green, but dorsum of head black, with the exception of a couple of minute green postocular spots.

Pronotum black. Synthorax mainly black, with irregular green bands. Legs black. Abdomen black dorsally, with narrow green end-rings of segments, sides greenish. In mature specimens, a blue pruinosity covers the face, dorsum of head, pronotum, sides and sternites of synthorax, the base or the legs and wings, and the sides of the abdomen. On the dorsum of the synthorax, this pruinosity produces two broad secondary antehumeral stripes. Appendages: superiors branched, the outer branch shorter and hooked inwardly at the apex. The outer branch extends as a finger-shaped projection well beyond the inner branch. App. inf. about half the length of app. sup., with broad base, from which arise two spoon-shaped leaflets. Accessory genitalia: lamina deeply but broadly cleft; hamuli as for genus.

Female

A much paler insect than the male, with reduced pruinosity except at the base of the wings. Face and dorsum of head brownish, with a dark stripe on the labrum, a fine black stripe on the frons, and some black spots near the ocelli and the antennal sockets. In old specimens, these individual black patches may merge into a single dorsal patch. Pronotum with hind margin depressed, smoothly rounded in the middle. Stylets short, not longer than the base of the hind lobe. Epaulettes present. Lam. mes. a broad flat, hollowed-out triangle. In the carinal fork appear two additional small triangles, adjacent to the lamina. Synthorax brown, with a black line on both sides of the carina. Carina itself brown. Synthorax brown, with a black line on both sides of the carina. Humeral black reduced to an anterior and a posterior spot. Abdomen brown, with broad black markings as in Fig. 154. Legs entirely brown. Wings as in male.

Measurements (mm): *Male.* Total length 35–38; abdomen 28–29. *Female.* Total length 34–37; abdomen 26–27.

Distribution: Endemic to the Levant, and distributed from northern Syria (Lake Homs: Gadeau de Kerville, 1926; Orontes north of Homs: Schneider, 1981a) to the Dead Sea, where it is still common in the wadis of the eastern slope of the basin.

Israel (Locality records): Jordan River at Benot Ya'aqov bridge (1), 'Ein Jalabina (1), Kefar Ruppin (7), Buteicha Swamps (7), near Tiberias (7), Wadi Fari'a (12), Aqua Bella (11), 'Ein Duyuk (13), Jericho (13), Wasit (18).

Pseudagrion torridum torridum Selys, 1876

Figs. 161–162, 165, 176–177, 183, 186

Pseudagrion torridum Selys, 1876:500. Andres, 1928:23; Ris & Schmidt, 1936:58; Pinhey, 1964:117.

Pseudagrion torridum torridum —. Dumont, 1973:186.

Type Locality: Dakar, Senegal.

Male

Mouth parts, face and dorsum of head green, marked with black as in Fig. 165. Postclypeus with trilobed black spot; frons with black band. Postocular spots large, confluent.

Synthorax green with broad carinal and humeral bands. Sides of synthorax changing to greenish-blue.

Wings: Pt light yellow.

Abdomen: S_{1-7} green, marked with black, sandglass-shaped spot. Appendages: app. sup. long, forcipate, with strong inner branch, separated from outer one by a deep ridge. Inner branch massive, sending two pointed projections ventrad, a distal stronger one, and a basal smaller one. App. inf. triangular in lateral view, shorter than app. sup. Legs light yellow, femora with black stripe. Accessory genitalia: lamina arched, converging into a point; hamuli as for genus. Ligula: L_2 rather elongate, with flanges extremely short.

Female

Face and head almost entirely brown, sparsely marked with black on anteclypeus, frons and ocellar area, and with a thin black bar between the compound eyes and the posterior ocelli. Pronotum orange-brown, hind lobe erect. Stylets well developed. Lam. mes. of complicated shape and relief, produced against the compressed carinal fork. Abdomen green with heavy black markings, contracted on S_{8-9}; S_{10} blue.

Measurements (mm): *Male.* Total length 30–34; abdomen 23–27. *Female.* Total length 30–34; abdomen 24–27.

Distribution: Most of Africa south of the Sahara, including the Nile Valley.

Sinai (Locality records): In Sinai, the species has been collected in the south-west: Abu Rudeis (23) and at Wadi Tayiba (22), Wadi Gharandal (22), Wadi Hibran (22) and Wadi Feiran (22). Capture data range from April to August.

Pseudagrion torridum hulae Dumont, 1973

Figs. 150–151, 159–160, 166–167, 178–179, 189

Pseudagrion torridum hulae Dumont, 1973:189.

Type Locality: Lake Hula, Israel.

The name applies to the northernmost population known; this subspecies is strikingly pale, paler than typical Afrotropical damseflies from Sinai.

Male

Black markings on head reduced (Fig. 166). Black stripes on synthorax narrow. Abdominal black markings, however, typical, but spot on S_{10} often fragmented. App. sup. with the distal edge of the inner branch more deeply hollowed-out than in the nominal subspecies.

Female

Black markings on the head reduced to very small spots (Fig. 167). Abdominal black markings more contracted than in the nominal subspecies.

Measurements (mm): *Male.* Total length 31–38; abdomen 25–26. *Female.* Total length 30–32; abdomen 24–24.5.

Distribution: Known only from Lake Hula (1) where it was collected in June.

Pseudagrion nubicum Selys, 1876

Figs. 155, 168, 174–175, 181

Pseudagrion nubicum Selys, 1876:501. Ris, 1912:159; Andres, 1928:23; Ris & Schmidt, 1936:24; Pinhey, 1964:92; Dumont, 1973:191.

Type Locality: "Nubia".

Male

Head greenish, frons partly black, dorsum of head black with blue spots near the ocelli, and rather large postocular spots, often confluent *inter se*.

Synthorax marked with black as in *P. torridum*. Antehumeral stripes green. Legs brown.

Wings with Pt dark brown.

Abdomen blue, heavily marked with black. S_2 with stalked U-shaped spot; S_{3-7} black; S_{8-9} blue; S_{10} black on dorsum. Appendages: app. sup. with outer and inner branch, separated by a ridge. Outer branch ending in an inwardly bent hook. Inner branch blunt-ending, with a series of small marginal teeth and a stronger spine near its base. App. inf. only slightly shorter than app. sup. Accessory genitalia: lamina broadly arched. Ligula: L_2 with short, rounded flanges.

Female

Labrum green, postclypeus green with basal black; rest of face orange-brown. Vertex and occiput with less black than in male. *Pronotum*: hind lobe somewhat convex in the middle, with long stylets and no epaulette. Lam. mes. very broadly triangular, impressed into the synthorax, the edges slightly raised outwardly. Carinal fork narrow with a rounded plate projecting from lower rim of margin. Antehumerals wider than half the width of the mesepisternum but variable. In young specimens, the humeral stripe is sometimes almost completely absent. Heavy black markings on abdomen. S_{10} blue, unmarked.

Measurements (mm): *Male.* Total length 30–32; abdomen 22.5–26. *Female.* Total length 29–33; abdomen 22–26.

Distribution: Widespread in the Afrotropical region, and occurring in the Nile Valley as well. The species is included here because Andres (1928) cites it from Suez. It might therefore well occur in Sinai.

Genus ERYTHROMMA Charpentier, 1840
Libellulinae eur., p. 148

Type Species: *Agrion najas* Hansemann, 1823.

Small to medium-sized damselflies, with hyaline wings. Pt similar in all four wings; apical reticulation of the wings dense, in the posteriors more so than in the anteriors. Body colours blue (males) or green (females) and black, with eyes bright red in living males. Dorsum of head without postocular spots or, rarely, punctiform spots. S_{10} in males not raised terminally. Female without vulvar spine. Legs robust, not flattened. d acutely pointed distally. Ab arising well distal to Ac, closer to level of an_1 than of an_2.

Erythromma viridulum orientale Schmidt, 1960
Figs. 190–194

Agrion viridulum Charpentier, 1840:149.
Erythromma viridulum —. Selys, 1876a:1304.
Erythromma viridulum orientale Schmidt, 1960:23. Dumont, 1977b:144.

Type Locality: El Rhab on Orontes River, Syria.

Male
Eyes, mouth parts, face red, but postclypeus black. Dorsum of head black without postocular spots. Two rounded red spots located in front of lateral ocelli.
Pronotum black, sides blue. Synthorax black down to halfway the mesepisternum. Antehumerals narrowing distally, and narrower than humeral stripes. Lateral synthoracic sutures black on azure blue base colour. Legs yellow, femora black on posterior surface.
Wings hyaline. Pt elongate, brown.
Abdomen: S_1, S_2, top of S_3 blue laterally; S_{3-7} black dorsally, greenish-yellow laterally; S_8 black on dorsum, blue on sides; S_9 totally black; S_{10} blue with sandglass-shaped black spot. Appendages: app. sup. as long as S_{10}, apically bifid in a ventral plane, the upper branch hooked inwardly. The whole appendix widened before its base, then constricted again. Inferiors very short, yellow, vertically oriented, with upturned black hook. Accessory genitalia as in *Coenagrion*.

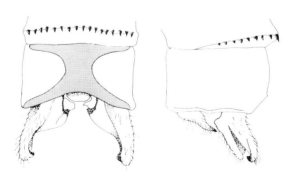

Fig. 190: *Erythromma viridulum orientale* Schmidt, 1960; male terminalia
dorsal and lateral views

113

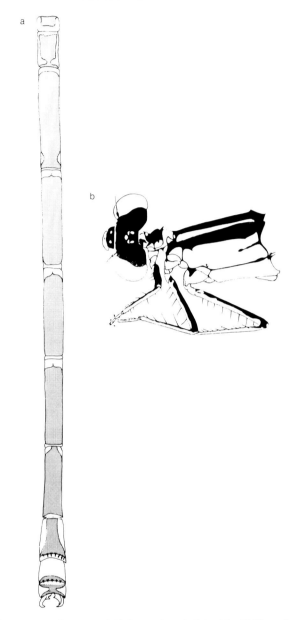

Fig. 191: *Erythromma viridulum orientale* Schmidt, 1960; male
a. abdominal markings; b. markings on head, synthorax and legs
(after Schneider, 1984)

Female

Head as in male, but clear parts yellow or green, not red. Pronotum black, sides yellow. Hind margin with upper ridge convex, upright, and lower margin angularly produced beyond it. Lam. mes. almost ellipsoidal, impressed into the synthorax, their bottom

114

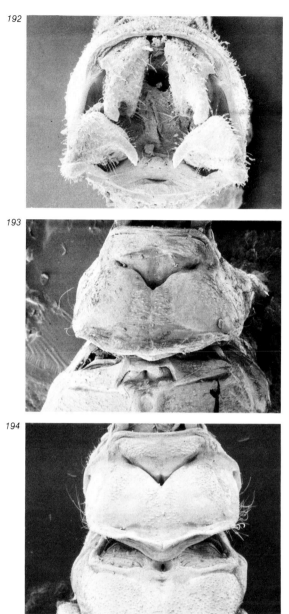

Figs. 192–194: *Erythromma viridulum orientale* Schmidt, 1960
192. male terminalia, posterior view;
193. male, pronotum and lamina mesostigmalis;
194. female, pronotum and lamina mesostigmalis

transversely ribbed. Pre-episternum curved along the external margin of the lamina. Carinal fork reduced to a central pit. Synthorax marked as in the male, but ground colour green, seldom blue. Abdomen blue or green, copiously marked with black.

Measurements: Two males from Lake Hula had a total length of 25.5 mm and 26 mm, abdomen 20 mm and 20.5 mm; specimens from Anatolia were in the range of 30–32 mm (total length), while the nominal subspecies in Central and Western Europe frequently exceeds 32 mm in total length.

Schmidt (1960) briefly characterized this subspecies as having wider antehumerals, and clear spots in front of the anterior ocelli. I find this a very relative and variable character and consider *E. viridulum orientale* a doubtful taxon, possibly a dwarf form, as is the case with *Coenagrion lindeni zernyi*.

Distribution: Macedonia, Greece, Anatolia, Iran, Iraq, Syria, Jordan, Israel. May–July.

Israel (Locality records): Lake Hula (1), Qiryat Shemona (1).

Genus CERIAGRION Selys, 1876
Bull. Acad. r. Belg., Ser. 2, 42:525

Type Species: *Agrion (Pyrrhosoma) cerino-rubellum* Brauer, 1865.

Damselflies of small to medium size. Colours non-metallic, red, orange or olivaceous, occasionally marked with black. Wings hyaline. Pt covering about one cell. d acutely pointed. Ab and Ac arising from the same point on the wing petiole or Ac arising from Ab distal to the point where Ab emerges from the wing margin. Head narrow; frons with a well-defined transverse ridge or crest. No postocular spots. Legs short. Anal appendages of males variously built. No vulvar spine in female. Accessory genitalia as for family.

Distribution: Palaearctic, Afrotropical and Oriental regions.

Key to the Species of Ceriagrion
(Figs. 195–203)

1. Wings: Ac arise from Ab, not from wing margin.
   Males: S_{10} raised into a semicircular tubercle, set with a crown of black spines. Dorsum of head and synthorax black; abdomen bright red.
   Female: head and thorax as in male, but abdomen red or red marked with bronze-black. Hind margin of pronotum an upright plate. Lam. mes. very small; carinal fork extremely wide with hind margins strongly raised in two upright, confluent triangular leaflets.
   Ceriagrion tenellum georgfreyi (Schmidt)

– Wings: Ac and Ab with a common origin on the wing margin.
   Males: S_{10} with a flat terminal tubercle, split into two widely separated lateral bosses, each set with a tuft of black spines. Dorsum of head and synthorax olivaceous-brownish; abdomen bright orange.
   Female: head, thorax and abdomen uniformly olivaceous-brownish. Hind margin of pronotum a depressed, rounded plate. Lam. mes. triangular, simple. Carinal fork shallow, as wide as one lamina, without a hind ridge.   **Ceriagrion glabrum** (Burmeister)

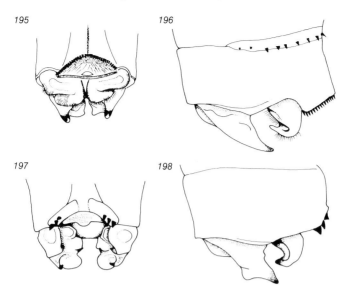

Figs. 195–198: *Ceriagrion* spp., male terminalia,
dorsal and lateral views
195–196. *C. tenellum georgfreyi* (Schmidt, 1953);
197–198. *C. glabrum* (Burmeister, 1839)

Ceriagrion tenellum georgfreyi (Schmidt, 1953)

Figs. 195–196, 199–201

Pyrrhosoma tenellum De Villers 1789:15. Selys, 1887:48; Morton, 1924:34.
Ceriagrion georg frey Schmidt, 1953:1. Schmidt, 1954a:83.
Ceriagrion tenellum georgfreyi —. Dumont, 1977b:144.

Type Locality: Sariseki near Iskenderun, Turkey.

Male
Mouth parts yellow; labrum, clypeus, genae, antennae, frons reddish. Dorsum of head entirely black; two comma-shaped yellow markings in front of and beside the lateral ocelli.
Pronotum black, its hind margin erect and somewhat concave in the middle. Synthorax black down to metathorax. Lam. mes. triangular, small. Carinal fork extremely wide and deep, its posterior margin forming a crest. Metathorax yellow, with Su_2 partly black, and a black patch on $epst_3$ and epm_3. Legs orange.
Wings: Ac arises from Ab, about midway between an_1 and an_2. Pt light brownish.
Abdomen uniformly red. S_{10} raised terminally into a tubercle, semicircular and hollowed-out in posterior view, fringed by a continuous crown of robust black spines. Appendages shorter than S_{10}; app. inf. somewhat longer than app. sup. in side view,

117

upturned over the superiors, pointed, and with a slight subapical swelling. App. sup. rounded, with inferior teeth and apical tuft of golden hairs. The appendix strongly hollowed-out beyond the lower spine; an upper spine is barely discernable. Accessory genitalia as in *Coenagrion*.

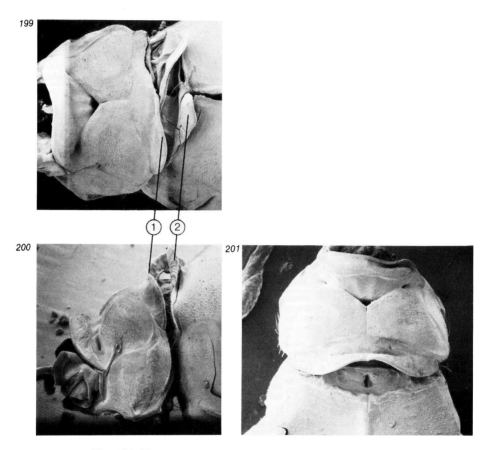

Figs. 199–201: *Ceriagrion tenellum georgfreyi* (Schmidt, 1953)
199. female, hind ridge of pronotum (1) and lamina mesostigmalis (2) in dorsal view;
200. the same, lateral view;
201. male, pronotum and lamina mesostigmalis, dorsal view

Female

Head as in male, but a black dot on labrum, and postclypeus and part of frons black. Pronotum black , with yellow or reddish spots on sides. Hind lobe erect. Lamina small and triangular, but carinal fork wide, its hind margin formed by a conspicuous triangular, upright lobe on each side, confluent with the lamina but not with each other. These two upright lobes, in lateral view, higher than the hind rim of the pronotum. Synthorax as in male, but humeral suture may be wholly or partly marked by a yellow or orange line. Sides of synthorax, wings, legs as in male. Abdomen entirely red, or partly marked with black dorsally. The black colours may be limited to S_{3-7}, but they may also cover the whole dorsum of the abdomen, in which case, the sides being yellow, no red colour is left at all. This last type, which is known in the nominal subspecies, has not yet been demonstrated to occur in Levantine or Anatolian populations.

Measurements (mm): *Male*. Total length 35–39; abdomen 28.5–31. *Female*. Total length 36–40; abdomen 29.5–33.

Distribution: The nominal subspecies and a third geographical subspecies (*C. tenellum nielseni*) inhabit Europe and the Maghreb countries of North Africa; *C. tenellum georgfreyi* is found in Anatolia, Syria, and almost certainly in the Lebanon and Jordan as well.

Israel (Locality records): Lake H̲ula (1), Jordan River near H̲ula (1), Binyamina (4), Bet She'an (7), Rosh Ha‘Ayin (8)

Ceriagrion glabrum (Burmeister, 1839)

Figs. 197–198, 202–203

Agrion glabrum Burmeister, 1839:821.
Ceriagrion glabrum —. Selys, 1876b:527; Andres, 1928:25; Longfield, 1932b:34.

Type Locality: Cape of Good Hope, S. Africa.

Male

Entirely devoid of black. Mouth parts and head uniformly yellow or orange, becoming darker in older specimens.

Pronotum, synthorax, legs all similarly coloured, the sutures slightly darker brown. Only the spines on the legs are black. Pronotum with hind margin a depressed, rounded plate. Lamina triangular; carinal fork shallow.

Wings: Ab and Ac rise from a single point on the petiole. Pt orange or brown.

Abdomen entirely bright orange. Hind rim of S_{10} raised but medially depressed, so that two lateral tubercles, each with a tuft of black spines, are formed. Appendages shorter than S_{10}, the inferiors longest, curved dorsad over superiors, pointed, with a rather strong subapical swelling. Superiors rounded, with a down-turned black spine. Accessory genitalia as in *Coenagrion*.

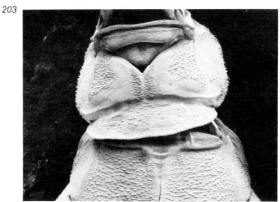

Figs. 202–203: *Ceriagrion glabrum* (Burmeister, 1839);
pronotum and lamina mesostigmalis
202. female; 203. male

Female

As the male, but all colours more brownish, sometimes olivaceous, and some sutures on the vertex and the synthorax may darken with time to the extent of appearing almost black. Internal tip of lam. mes. black. Abdomen dark brownish-orange, the dorsum of the terminal segments sometimes very dark. Legs and wings as in male.

Measurements (mm): Male. Total length 37–43; abdomen 29–34. *Female.* Total length 41–45; abdomen 32–36.

Distribution: Africa south of the Sahara. Specimens have been found near Cairo, and in Kharga Oasis (Andres, 1928). The possibility exists that the species occurs in oases in Sinai.

Suborder ANISOPTERA

Key to the Families of Anisoptera

1. Compound eyes completely and rather widely separated.   **Gomphidae**
- Eyes meeting across dorsum of head, either in one point, or over a longer distance   2
2. d similar in all wings, its costal edge much longer than its proximal edge. Females with ovipositor. Males with oreillettes and with hind wings angulate at their base   3
- d dissimilar in forewings and hind wings, situated well distal to arc in forewing, at or adjacent to arc in hind wing. Females without ovipositor, usually with a pair of smallish vulvar valvules. Males without oreillettes, and with hind wings rounded at their base.   **Libellulidae**
3. Eyes meeting in a point only. Thorax and abdomen coloured yellow and black in both sexes. Females with long, straight, blunt ovipositor that extends well beyond the tip of the abdomen.   **Cordulegasteridae**
- Eyes broadly contiguous. Coloration variable according to species and sex, usually a mosaic of brown, green or blue, yellow, but never yellow-and-black. Females with an ovipositor of zygopterous structure, not extending beyond tip of abdomen.   **Aeschnidae**

Family GOMPHIDAE

Medium-sized, rarely large dragonflies, coloured in various tinges of yellow, sometimes ochraceous or brown, and black, rarely pruinose. Head with eyes rather widely separated, and the vertex often with diagnostic modifications. Wings hyaline, sometimes slightly smoky at their base. Sectors of arc separated from their origin. Membranula very small, almost absent. Pterostigma elongate, medium-sized, sometimes dilated in the middle. d rather similar in fore wings and hind wings, but its longest axis vertical in fore wing, horizontal in hind wing. Hind wing of males excavated and angulate at the base, in females rounded. Antenodals always numerous (9–18), and primaries always readily visible. Male appendages variously structured, the inferiors always somewhat bifid or notched, frequently deeply cleft. Accessory genitalia genus-specific, often even species-specific. Females without ovipositor, with a pair of smallish vulvar valvules on S_8.

The oriental genus *Anormogomphus*, found in Iraq, differs, among other things, by the fact that the base of the hind wing is rounded in both sexes (Fig. 207). *Anormogomphus kiritchenkoi* Bartenef is not likely to occur in the Levant.

121

Key to the Genera of Gomphidae
(Figs. 2, 10, 29, 204–297)

1. Wings: both triangle (d) and hypertriangle (ht) traversed. Membranula well developed, dark coloured. **Lindenia** Selys
–   Wings: d and ht entire. Membranula reduced 2
2. Males 3
–   Females 5
3. S_8 and S_9 with foliate expansions. Superior appendages closely apposed, curved in a dorso-ventral plane. Inferior appendages at most half the length of the superiors. **Paragomphus** Cowley
– S_8 and S_9 without foliate expansions. Superior appendages separated at their base, often divaricate. Inferior appendages more than half the length of the superiors 4
4. Superior appendages form a forceps. Inferior appendages rather narrow, deeply cleft, and always parallel. **Onychogomphus** Selys
–   Superior appendages divaricate, pointed. Inferior appendages notched, divaricate. **Gomphus** Leach
5. Sternite of S_9 without particular modifications around and distal to vulvar scales, at most transversely ribbed. **Gomphus** Leach
–   Sternite of S_9 with a well-circumscribed, rounded or squared membranaceous area, and/or with median crests 6
6. S_9 with a semicircular or angular field behind the valvules, but without median crests. Hind wing: at least two sectors descend directly from the anal vein to the wing margin between its base and the distal angle of d. **Paragomphus** Cowley
– S_9 with a more or less well-defined semicircular field between the valvules, and with median sclerotized crests. Hind wing: cells subjacent to anal vein, between its base and the distal edge of d, organized as a closed anal field of 2–3 cells or at least so different in shape from lower cells that no direct vertical sectors between the anal vein and the wing margin are formed. **Onychogomphus** Selys

Genus LINDENIA Selys, 1840
Monogr. Libellul. Eur., p. 74

Type Species: *Aeshna tetraphylla* Vander Linden, 1825.
Large gomphids. Wings with d composed of 2–4 cells; an anal field of 3–5 cells in hind wing. Membranula present, large, dark brown. Pt long. Wings sometimes suffused with amber. Legs black with yellow stripes, robust. Synthorax partly pruinose. Abdominal segments 7 and 8 with lateral foliate expansions in both sexes. Males: app. sup. filiform; inferiors short, triangular. Female with vulvar scales short, forked. No particular modifications on sternite of S_9.
Distribution: S.E. Europe, N. Africa, and the Irano-Turanian area (see further).
One regional species. A little known second species (*L. inkiti* Bart, 1929) inhabits the Caucasus.

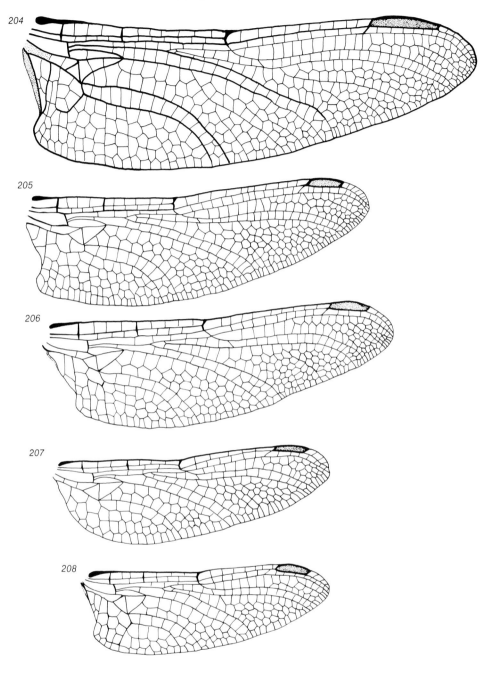

Figs. 204–208: Male gomphid wings
204. *Lindenia tetraphylla* (Vander Linden, 1825);
205. *Gomphus* sp.; 206. *Onychogomphus* sp.;
207. *Anormogomphus kiritchenkoi* Bartenef; 208. *Paragomphus* sp.

Lindenia tetraphylla (Vander Linden, 1825)
Figs. 204, 209–214

Aeshna tetraphylla Vander Linden, 1825:32.
Lindenia tetraphylla —. Selys, 1840:76. Selys & Hagen, 1850:102; Selys, 1887:31; Waterston, 1980:61; Schneider, 1981b:97.
Vanderia tetraphylla —. Kirby, 1890:78.

Type Locality: Lake Averno near Naples, Italy.

Male
Mouth parts yellow; labrum, clypeus, frons greenish. In aged specimens, clypeus and frons darkened. Frons with a basal black stripe. Vertex black around ocelli and with strongly raised, bilobed brownish or olivaceous crest. Occiput and pronotum black.
Synthorax green-yellow, very heavily marked with black anteriorly. In young specimens, carina and an obliquely placed lunule yellow. In aged males, front of

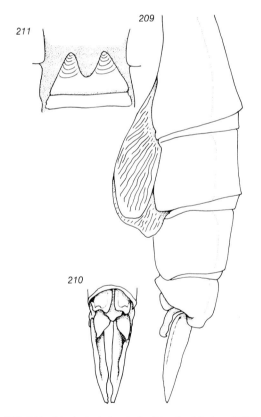

Figs. 209–211: *Lindenia tetraphylla* (Vander Linden, 1825)
209. male terminalia, lateral view; 210. the same, ventral view;
211. female, vulvar valvules

synthorax entirely black. Sides greenish with broad black bands, covered by blue pruinosity. Legs black with yellow stripe on femora.

Wings: costa yellow, rest of venation black. Pt long, brown with thick black margins. Membranula present on both pairs of wings, brown, with paler margins. d traversed, 3–5 celled. Some basal amber, especially in cubital space of hind wing, the latter deeply angulate at its base. Anal field well defined, 3–5 celled. In old specimens, the whole wing sometimes suffused with amber.

Abdomen uniformly black in old specimens. In tenerals an oblong yellow median patch extends on the dorsum of S_{2-7}. Sides of S_{1-2} and S_{7-10} brownish-yellow. Oreillettes with denticles. Foliate espansions brown, directed posteriorly. Appendages: superiors about twice as long as S_{10}, straight. Inferiors shorter than S_{10}, triangular. Accessory genitalia: lam. ant. shallowly excavated; ham. ant. pointed, with subapical spine; ham. post. complex (Fig. 213) larger. Vesica small, inconspicuous, hidden under ham. post.

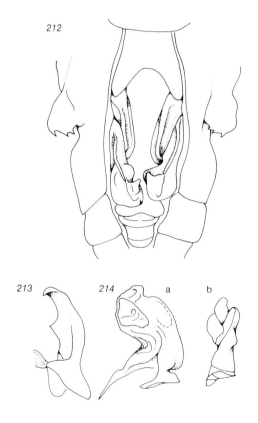

Figs. 212–214: *Lindenia tetraphylla* (Vander Linden, 1825); male, accessory genitalia

212. ventral view; 213. hamulus 1; 214. a–b. hamulus 2

125

Female

As male, but clearer, and abdomen not constricted. Dorsum of S_2 with median yellow streak; S_3 clear brown with black markings near endring; a conspicuous white spot on its flanks. S_{3-7} black, with dark yellow markings. S_{8-10} black with clear lateral spots. Foliate expansions brown. Styli dark and pointed, longer than S_{10}. Vulvar scales simple, short.

Measurements (mm): *Male*. Total length 69–80; abdomen 49–57. *Female*. Total length 70–75; abdomen 50–55.

Distribution: A lake species, on the wing between May and July, found in Italy, Yugoslavia, Albania, Greece, the Caucasus(?), and Transcaucasian states of the U.S.S.R., the Caspian Basin, Iran, Afghanistan, Baluchistan, Iraq, Saudi Arabia, the Levant, Egypt. In the Maghreb, it has been found with certainty in N.E. Algeria only. Recently, a migratory movement of this species, involving great numbers of specimens, was reported from Jordan by Schneider (1981).

Israel (Locality records): Ḥula (1), Gonén (1), Sedé Neḥemya (1), Karé Deshe (7), Gesher (7), Deganya (7).

Genus GOMPHUS Leach, 1815
Edinburgh Encyclopaedia, 9:137

Type Species: *Libellula vulgatissima* Linnaeus, 1758.

Size moderate. Colour yellow marked with black. Head robust, frons angulate, never with a median depression; occiput simple. No oreillettes on rear of head. Wings with dense reticulation, tornus angulate in male, membranula reduced. Anal lip absent. d entire. Pt of moderate length, usually somewhat swollen in the middle. Appendages simple, the superiors divaricate, apically pointed. Inferiors broadly but shallowly bifid, their two branches divaricate like the superiors. Vesica spermalis with a very strongly developed basal segment. Female: valvules simple, the sternite of S_9 not modified by sclerotized crests, at most transversely ribbed.

Distribution: Europe, N. Africa, Asia, N. America

Key to the Species of Gomphus
(Figs. 205, 215–248)

1.   Black antehumeral and carinal stripes merge in front and posteriorly to enclose an elongated, ellipsoidal yellow spot.   **Gomphus flavipes lineatus** Bartenef

–   Black antehumeral and carinal stripes run parallel and do not merge to isolate a frontal yellow spot on the synthorax   2

2.   Legs largely yellow.

    Male: S_{10} with mid-dorsum at least partly yellow. Ham. post. strongly developed, their top rounded and flattened, but without a strong anterior hook. Basal segment of vesica spermalis very strongly developed, bulb-shaped, yellow with black base-line.

126

Female with continuous mid-dorsal yellow streaks on abdomen. Vulvar scales short and blunt apically. Their base on S_8 with a lateral swelling on either side. Vertex with a horn behind each lateral ocellus. **Gomphus davidi** Selys

– Legs largely black.

Male: S_{10} black. Ham. post. with strong anteriorly curved apical hook and interior swelling. Basal segment of vesica spermalis only moderately swollen, black.

Female: a discontinuous black line runs across the mid-dorsum of the abdomen, particularly on S_{8-10}. Vulvar scales rather short and blunt, no swellings at their base. Vertex without horn behind each lateral ocellus.

Gomphus vulgatissimus schneideri (Selys & Hagen)

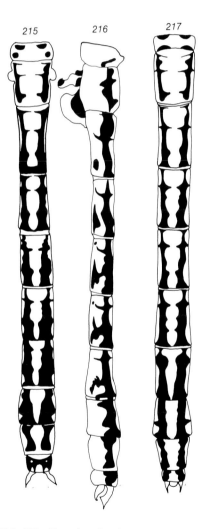

Figs. 215–217: *Gomphus davidi* Selys, 1887; abdomen
215. male, dorsal; 216. male, lateral; 217. female, dorsal

218

219

Figs. 218–219: *Gomphus davidi* Selys, 1887; synthorax
218. male; 219. female

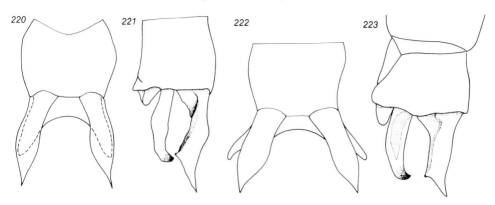

Figs. 220–223: *Gomphus* spp., male appendages,
dorsal and lateral views
220–221. *G. davidi* Selys, 1887;
222–223. *G. vulgatissimus schneideri* (Selys & Hagen, 1850)

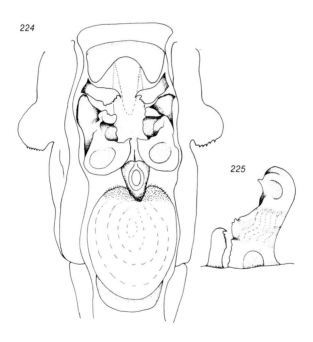

Figs. 224–225: *Gomphus davidi* Selys. 1887; male
224. accessory genitalia; 225. hamuli, lateral view

129

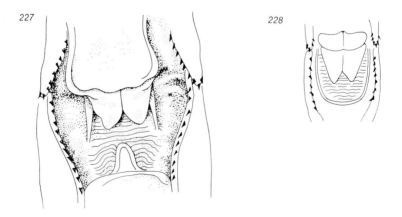

Figs. 226–228: *Gomphus* spp., females, vulvar scales
226. *G. davidi* Selys, 1887;
227. *G. flavipes lineatus* Bartenef, 1929;
228. *G. vulgatissimus schneideri* (Selys & Hagen, 1850)

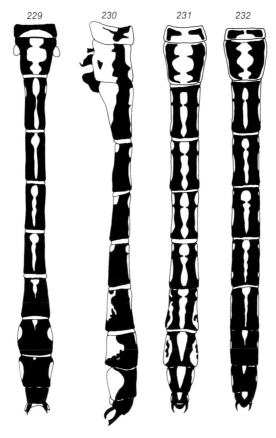

Figs. 229–232: *Gomphus vulgatissimus schneideri* (Selys & Hagen, 1850);
abdominal markings
229–230. male, dorsal and lateral views;
231–232. female, dorsal and lateral views

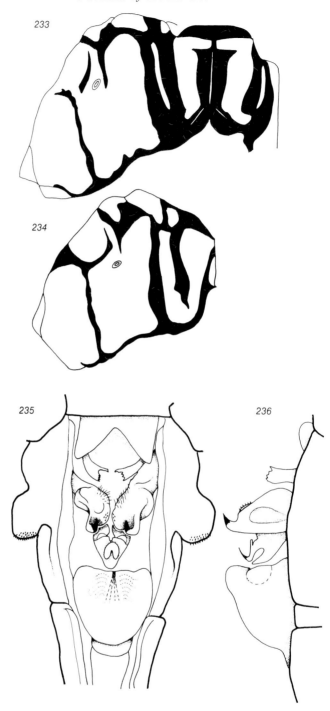

Figs. 233–236: *Gomphus vulgatissimus schneideri* (Selys & Hagen, 1850)
233. male; synthorax; 234. female, synthorax
235–236. male, accessory genitalia, ventral and lateral views

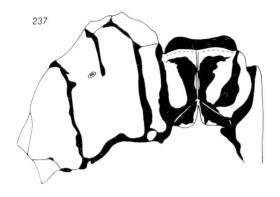

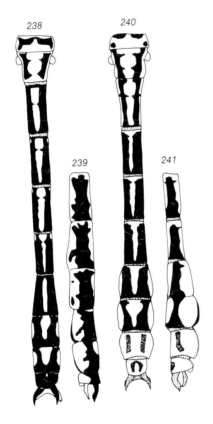

Figs. 237–241: *Gomphus flavipes*, males
237. synthorax;
238–239. *G. flavipes flavipes* (Charpentier, 1840),
abdomen, dorsal and lateral views;
240–241. *G. flavipes lineatus* Bartenef, 1929;
abdomen, dorsal and lateral views

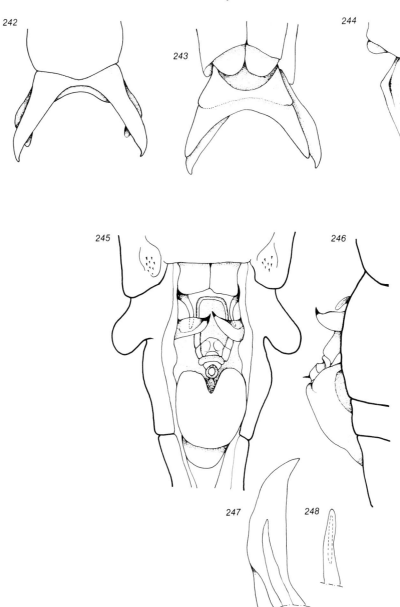

Figs. 242–248: *Gomphus flavipes lineatus* Bartenef, 1929; male,
terminalia, dorsal, lateral and ventral views
245–246. accessory genitalia, ventral and lateral views;
247. hamulus 2; 248. hamulus 1

Gomphus davidi Selys, 1887

Figs. 215–221, 224–226

Gomphus davidi Selys, 1887:30. Morton, 1924:40; Gadeau de Kerville, 1926:79; Schmidt, 1938:147; St. Quentin, 1964c:49; St. Quentin, 1965:538; Dumont, 1977b:145; Schneider, 1981a:138.

Type Locality: Damascus, Syria.

Male

Mouth parts, face anteriorly yellow. A narrow black line between frons and clypeus. Base of frons, vertex black, occiput yellow.

Pronotum yellow, marked with black. Synthorax yellow, with broad carinal black stripe, broad antehumeral black band about midway between carinal and humeral stripe, and black stripe on humeral suture, about as wide as or slightly narrower than antehumeral stripe. Lateral synthoracic sutures rather narrowly marked with black. Legs yellow, with black stripes on the tibiae, and on the femora of the first and second pairs. Femur of the third pair entirely yellow, or with finest black stripe.

Wings as for genus. Pt light brown.

Abdomen bright yellow marked with black. Mid-dorsal yellow stripes continuous over the whole length of the abdomen. S_{10} always with a mid-dorsal black spot, rest yellow. Appendages: superiors rather long and finely pointed, apex with inner margin straight, but outer margin concave, so that the tips of the appendix are slightly divaricate. In side view the appendix is angulate at the level of its constriction. App. inf. deeply and widely emarginate, apex upturned. App. sup. yellow in tenerals, turning brown in later life. Accessory genitalia: lam. ant. moderately emarginate; ham. ant. small, black, scoop-shaped, indented; ham. post. strongly developed, with broad base, and a flattened, rounded apex with a small tooth. Vesica spermalis with a very strongly developed sperm reservoir, largely yellow coloured, with sides partly black, somewhat hollowed-out anteriorly to accommodate the backwardly curved apical segment of the vesica, set with stout golden hairs around the excavation.

Female

Head coloured as in male. Vertex with two outwardly pointing horns behind the lateral ocelli. Occiput wider than in male. Synthoracic black stripes somewhat narrower than in male. Wings, legs as in male. Abdomen yellow marked with black, but a continuous mid-dorsal, broad yellow stripe is present along its entire length. Styli yellow. Vulvar scales short, with broadly rounded tips. Their base on S_8 with a swelling on either side.

Measurements (mm): Male. Total length 48–55; abdomen 35–39. *Female.* Total length 46–55; abdomen 34–40.

Distribution: The south-east Mediterranean coast of Turkey, Syria, the Lebanon, Jordan, Israel. This is possibly the only true *Gomphus* that occurs in the Jordan Valley. Capture dates range between March and June. A running-water species.

Israel (Locality records): Hulata (1), Lake Hula (1), 'Ein Jalabina (1), Wadi Fari'a (6), 'Ubeidiya (7), Bet Yerah (7), Binyamina (8).

Gomphus vulgatissumus schneideri (Selys & Hagen, 1850)
Figs. 222–223, 228–236

Gomphus forcipatus, non *Onychogomphus forcipatus* (L). Schneider, 1845:114.
Gomphus schneiderii Selys & Hagen, 1850:293. Morton, 1915b:130.
Gomphus schneiderii race of *G. vulgatissimus*. Selys & Hagen, 1857:132.
Gomphus vulgatissimus race of *G. schneiderii* —. Selys, 1887:29.
Gomphus simillimus Selys, 1840:85. Kempny, 1908:264; Gadeau de Kerville, 1926:79.
Gomphus schneideri —. Schmidt, 1954a:59; St. Quentin, 1965:538; St. Quentin, 1968:493; Dumont, 1977b:144.
Gomphus vulgatissimus schneideri —. St. Quentin, 1964a:423.

Type Locality: Kellemisch (Gelemish), western Anatolia, Turkey.

Male
Mouth parts yellow; labrum fringed with black and with black median virgule. Clypeus yellow with black margin. A basal black stripe at suture between frons and clypeus. Frons yellow, vertex black, occiput yellow. Legs black. Pt dark brown, almost black. Synthorax yellow, marked with black. Antehumeral band wide, frequently confluent with broad humeral band near its upper end. Remaining lateral synthoracic sutures more narrowly marked with black.
Abdomen black, marked with mid-dorsal yellow stripes. These spots, however, not continuous across the different segments, but interrupted near the base and end-rings and reduced to a basal elongated spot on S_{7-9}; S_{10} and sometimes also S_9 uniformly black. Well-developed lateral yellow markings occur on S_{1-3} and S_{7-10}. Appendages black. Apex of superiors abruptly constricted; external and internal margins with about the same degree of concavity. In side view, an angulation at the level of the constriction. Inferiors hollowed-out, with upturned apex. Genitalia entirely black. Lamina deeply hollowed-out, with pointed invagination. Ham. ant. small, apically serrated. Ham. post. strong, with broad base, internal margin swollen and produced, and external apex a strong, anteriorly turned hook. Vesica spermalis with semicircular sperm reservoir, truncated anteriorly, notched apically, black coloured.
Female
Colours on head and thorax as in male. Vertex hollowed-out behind the ocelli, with two lateral tubercles but no horns. Abdomen yellow, copiously marked with black on the sides. Mid-dorsal yellow stripe continuous or almost so on S_{1-7}, more reduced on S_{8-10}. Legs black, but femora partly yellow. Vulvar scales rather long, about 1/2 of S_9, broad, with rounded apex and no swellings at their base on S_8.
Measurements (mm): *Male.* Total length 41–47; abdomen 30–34. *Female.* Total length 40–47; abdomen 30–33.
Distribution: A spring and early summer species, found in Greece, Anatolia, the Caucasus, Iran, northern Iraq. Also recorded, under *G. simillimus*, from Lake Homs, Syria, and probably still occurring in the Lebanon. Not yet found in the Jordan Valley. This might be another species limited on its southern boundary by the Nehring line.

136

Gomphus flavipes lineatus Bartenef, 1929

Figs. 227, 240–248

Gomphus flavipes (Charpentier, 1825:24). Martin, 1912:6.
Gomphus Davidi Selys 1887:30. Martin, 1912:6.
Gomphus f. lavipes var. *lineatus* Bartenef, 1929:61 (*lapsus calami*).
Gomphus Ubadschii Schmidt, 1953:6.
Gomphus ubadschii —. Schmidt, 1954a:60; Schmidt, 1954b:247; Sage, 1960a:121.
Stylurus (Gomphus) lineatus —. Schmidt, 1961:412.
Stylurus ubadschii —. Asahina, 1973:23.
Gomphus flavipes lineatus —. St. Quentin, 1965:538; Dumont, 1977b:145.

Type Locality: Poti, Georgian S.S.R.

Male

Mouth parts, genae, labrum, clypeus, frons yellow or green-yellow. A robust black stripe present between frons and clypeus, and at base of frons. Vertex greenish behind ocelli, separated from greenish occiput by a black transverse bar.

Pronotum yellow and black. Synthorax: carina yellow, flanked by a humeral stripe which widens anteriorly, and narrowly or broadly fuses both here and near the alar sinus with the black antehumeral stripe, to enclose an elongate yellow or greenish spot. Antehumeral stripe equidistant from carinal and humeral black stripes. Lateral synthoracic black markings narrow, green-yellow colour dominant. Legs black, flexor side of femora yellow.

Wings hyaline, Pt brown.

Abdomen: a fairly continuous dorsal yellow stripe, flanked by broad black bands, on S_{2-8}. Black stripes contract on S_{7-8}, so that sides of S_{8-10} are largely citron yellow. Dorsum of S_{10}, to a minor degree of S_9, yellow, with reduced brown markings. App. sup. brown, forcipate, finely pointed. App. inf. deeply emarginate, with yellow base and dark brown apices. Genitalia: lam. ant. squared, very narrowly and shallowly cleft; ham. ant. small, blade-shaped, tip rounded; ham. post. longer, knife-shaped, acutely pointed, curved anteriad. Vesica spermalis with large, yellow-coloured sperm reservoir. In side view, a black spot at the base of the reservoir.

Female

Head as in male, but black stripe between clypeus and frons much narrower. Vertex hollowed-out in its centre, and with two lateral tubercles behind the lateral ocelli. Synthorax, legs, wings as in male. Abdomen with a continuous yellow stripe on the dorsum, broadened to a wide spot on S_9. S_{10} yellow with two minute brown dots at the base of the segment. Sides of S_{7-10} bright citron yellow. Vulvar scales less than half the length of S_9, broad-based, massive, with obtuse apex. Wing base sometimes slightly suffused with amber.

Measurements (mm): *Male.* Total length 44–49; abdomen 32–36. *Female.* Total length 47–51; abdomen 34–36.

Distribution: From the south-west coast of Anatolia to Iran and northern Iraq, and southwards extending into Syria. Probably also in the Lebanon and parts of Jordan. Not yet found in the Jordan Valley. A riverine species, with capture dates ranging from May till August.

Genus ONYCHOGOMPHUS Selys, 1854
Bull. Acad. r. Belg., Ser. 2, 21:30

Type Species: *Libellula forcipata* Linnaeus, 1758.

Size medium, colour yellow marked with black and brown. Male with long, curved distinctly forcipate superior appendages; inferior appendages narrow, deeply bifid, closely apposed. Wings with tornus angulate; an anal field of 2–3 cells is present in most species. In females, and in males of *O. flexuosus*, this may be less distinct, and consist of cells only slightly larger than the underlying ones. d entire. Legs short. Male genitalia: vesica with basal segment never deeply hollowed-out. Ham. ant. forked. Ham. post. variously shaped, pointed. Females with vulvar scales rather short, pointed or rounded, sometimes with sclerotized swellings at their base, and with sclerotized ridges on the floor of S_9 behind the valves.

Distribution: Africa, Europe, and Asia.

Three species are regional, all with comparatively narrow ranges. Two more (one of which comprises at least three different subspecies) occur in Anatolia.

Note

Selys, in his various writings, has not associated the correct females with the males of three regional species. In particular, Selys' type of *Onychogomphus macrodon* female is a female of *O. lefebvrei*; the keys and descriptions offered here have therefore been based on new topotypical material.

Key to the Species of Onychogomphus
(Figs. 206, 249–277)

1.	Males	2
–	Females	4
2.	App. inf. as long as app. sup., simple, without swellings or inflexions. App. sup. curved inwards almost at right angles slightly more than half their length. A central "finger" at the level of the inward bend. Sperm reservoir of the vesica not modified to accommodate its own top segment, i.e. simply overlying the reservoir.	

Onychogomphus lefebvrei (Rambur)

–	App. inf. shorter than app. sup., and either with a deep inflexion or with a swelling. App. sup. gently curved inwards well apical to their middle, and without a "finger". Sperm reservoir of vesica shallowly excavated to accommodate its own top segment	3
3.	App. sup. with a single subapical spine. App. inf. S-shaped, with a lateral spine at the level of their inflexion. S_{7-9} without well-defined central black markings, at most with diffuse	

138

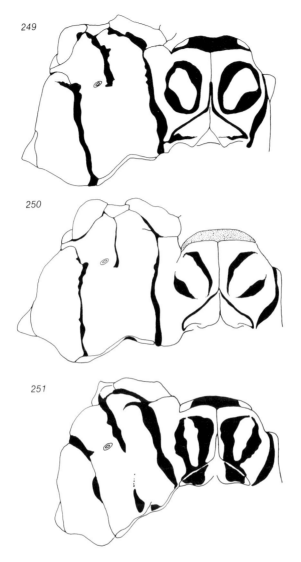

Figs. 249–251: *Onychogomphus* spp.
249–250. *O. lefebvrei* (Rambur, 1842); synthorax, male and female;
251. *O. flexuosus* (Schneider, 1845); male synthorax

brown dots. Sperm reservoir erect with apical U-shaped invagination to accommodate tip of its apical segment; the latter long and narrow.

Onychogomphus flexuosus (Schneider)

App. inf. with a strong swelling; their tips pointed, but without apical spine. App. sup. with an external row of spines between the level of their inward bend and close to their tip (the row, however, not reaching the top of the appendix). S_{7-9} with a mid-dorsal black spot. Reservoir of vesica spermalis erect, with two wing-shaped outgrowths on top, against which its broad and short apical segment rests.

Onychogomphus macrodon Selys

139

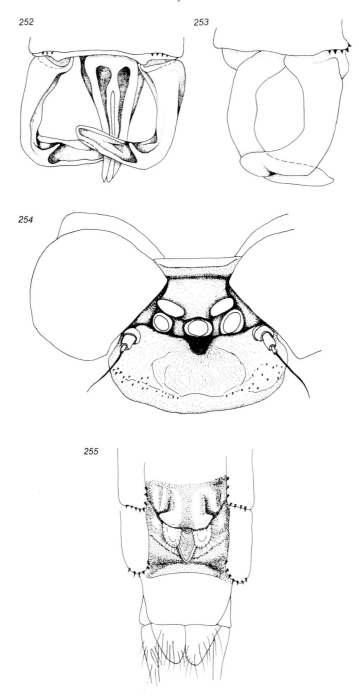

Figs. 252–255: *Onychogomphus lefebvrei* (Rambur, 1842)
252–253. male, terminalia, dorsal and lateral views
254. female, head; 255. female, valvulae

4.   Oreillettes on rear of head; a very deep hole at the base of the frons, in front of the anterior ocellus.   **Onychogomphus lefebvrei** (Rambur)

–   No oreillettes on rear of head; base of frons in front of anterior ocellus at most somewhat depressed   5

5.   Strong black markings on dorsum of S_{7-9}; a shallow depression in front of the anterior ocellus; occiput wide and short; floor of S_9 in front of vulvar scales with two V-shaped crests, connected by a mid-ventral sclerotized ridge.   **Onychogomphus macrodon** Selys

–   No strong black markings on dorsum of S_{7-9}, at most some diffuse brown spots. No depression on frons in front of anterior ocellus; occiput with two V-shaped crests that converge in one point.   **Onychogomphus flexuosus** (Schneider)

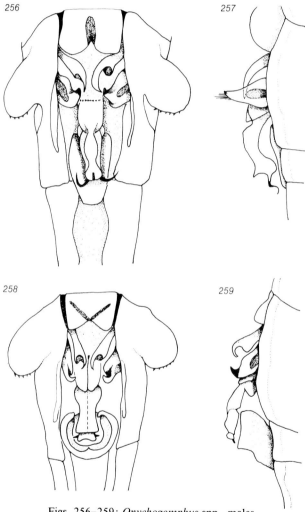

Figs. 256–259: *Onychogomphus* spp., males,
accessory genitalia, ventral and lateral views
256–257. *O. lefebvrei* (Rambur, 1842);
258–259. *O. flexuosus* (Schneider, 1845)

141

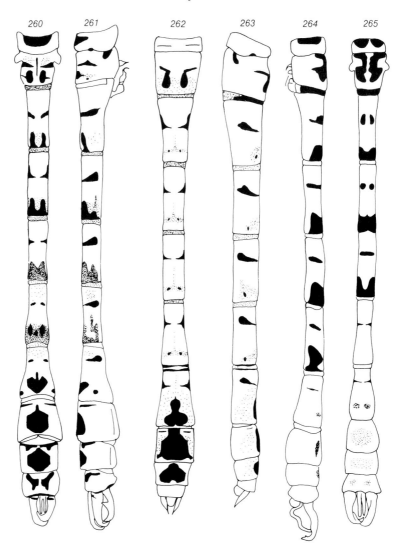

Figs. 260–265: *Onychogomphus* spp.; abdominal markings
260–261. *O. macrodon* Selys, 1887; male, dorsal and lateral views;
262–263. *O. macrodon*, female, dorsal and lateral views;
264–265. *O. flexuosus* (Schneider, 1845); male, dorsal and lateral views

142

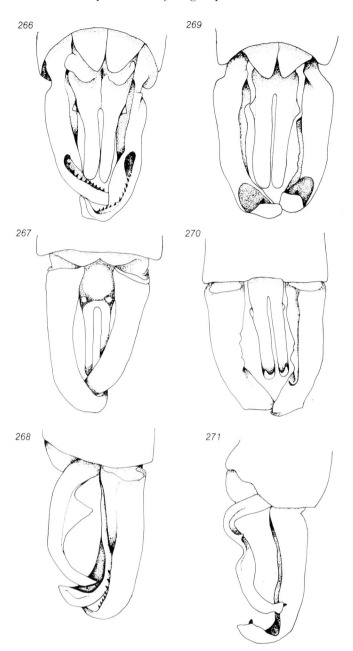

Figs. 266–271: *Onychogomphus* spp., male,
terminalia, ventral, dorsal, and lateral views
266–268. *O. macrodon* Selys, 1887;
269–271. *O. flexuosus* (Schneider, 1845)

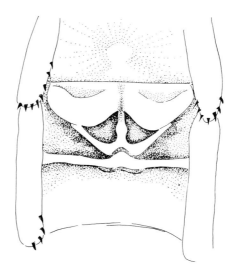

Fig. 272: *Onychogomphus flexuosus* (Schneider, 1845): female, valvulae

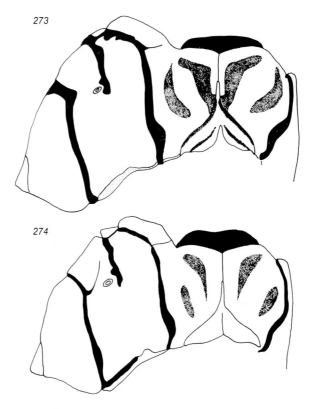

Figs. 273–274: *Onychogomphus macrodon* Selys, 1887; synthorax, male and female

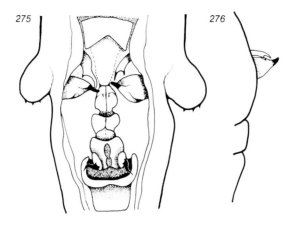

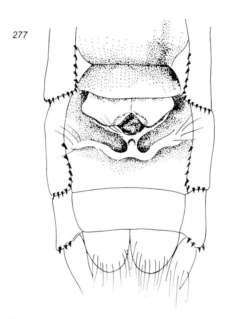

Figs. 275–277. *Onychogomphus macrodon* Selys, 1887
275–276. male, accessory genitalia, ventral and lateral views
277. female, valvulae

Onychogomphus lefebvrei (Rambur, 1842)

Figs. 249–250, 252–257

Gomphus lefebvrei Rambur, 1842:166. Selys & Hagen, 1854:37.
Gomphus forcipatus race *lefebvrei* —. Selys & Hagen, 1857:33.
Onychogomphus lefebvrei —. Selys, 1887:27; Morton, 1924:39; Andres, 1928:27; Schmidt, 1954a:59; Asahina, 1974:107; Dumont, 1977b:147.
Onychogomphus forcipatus lefebvrei —. St. Quentin, 1965:539.

Type Locality: Oasis of Bahrieh, Egypt.

Male

Mouth parts yellow. Labrum green-yellow with very thin black base. Frons green-yellow. Suture between frons and clypeus black in old specimens. Base of frons and ocellar area black. Vertex green-yellow. Occiput yellow.

Synthorax yellow. Carinal black stripes separated, usually forming a closed loop with the antehumerals. This loop may, however, be open anteriorly, and reduced to two narrow stripes in "pale" populations. Lateral sutures narrowly black. Legs yellow with a black spot at the top of each femur, and a black stripe on the tibiae.

Wings: Pt light brown. Anal field composed of two cells.

Abdomen yellow, marked with black near the end-rings of segments 2–7. Apical segments slightly widened, dark yellow or brownish, with diffuse dark brown, rarely black markings. Superior appendages forcipate, strongly bent inwards slightly more than half their length; their tip spatulate but with an upper finger-like expansion near the level of the inward bend. Inferiors about as long as the superiors, narrow and deeply cleft, curved upwards. Accessory genitalia: lam. ant. swollen, slightly invaginated; ham. ant. well developed, deeply forked; ham. post. erect, triangular, with a fine internal apical point, turned anteriorly, and a tuft of long yellow hairs. Vesica spermalis with sperm reservoir flat, flask-shaped; the elongated apical segment ("glans") simply overlying it. Apical flagellae of glans very long.

Female

Head entirely yellow to base of frons. Ocellar area black. Vertex yellow, with two lateral swellings. Occiput yellow. Rear of eyes with a narrow medial black band, rest yellow. A pair of yellow oreillettes present on the back of the eyes, lateral to the occiput. The frons, immediatly in front of the anterior ocellus, with a very conspicuous, deep hole (in which the apices of the app. sup. of the male are accommodated during tandem formation and copula). Legs, wings, and synthorax as in male. Abdomen cylindrical throughout, yellow marked with black. Vulvar scales strongly developed, rounded, with hairy margins, and with a deep and broad invagination almost down to their base. Base produced into tubercles. Floor of S_9 with two ridges, U-shaped, broadly touching each other medially, the inner, narrower ridge slightly but consistently protruding.

Measurements (mm): *Male.* Total length 46–52; abdomen 35–40. *Female.* Total length 45–50; abdomen 33–37.

146

Distribution: Anatolia, Iran, Afghanistan, Iraq, Syria, the Lebanon, Jordan, Israel, and Egypt. While the type locality is in Egypt, no recent records from that country have become available. In Israel, the species has been found between April and June, mainly in the northern part of the country.

Israel (Locality records): Wadi Qurein (1), 'Ein Jalabina (1), Banias River (1), Mt Tabor (2), Naḥal Daliyya (3), Wadi 'Auja (13).

Onychogomphus flexuosus (Schneider, 1845)

Figs. 251, 258–259, 264–265, 269–272

Gomphus flexuosus Schneider, 1845:114. Selys & Hagen, 1850:295.
Onychogomphus flexuosus —. Selys, 1887:27; St. Quentin, 1965:538; Dumont, 1977b:148.

Type Locality: Kellemisch (Gelemish), western Anatolia, Turkey.

Male

Mouth parts yellow; front of frons diffusely black. Ocellar region black. Vertex yellow, lateral tubercles more depressed than in *O. lefebvrei*. Occiput yellow, flat, narrower but longer than that in *O. lefebvrei*.

Dorsum of synthorax with carinal and antehumeral stripes forming a closed, narrow loop that may be narrowly open anteriorly. Lateral sutures marked with black on a greenish background. Legs yellow with a black patch on top of the femora. Tibiae yellow.

Wings: Pt light brown with thick black margins. Anal field not well defined, but usually cells immediately subjacent to anal vein larger than lower-level cell.

Abdomen straw-yellow, marked with black near the end-rings and the top third of S_{3-6}; apical segments widened, with diffuse markings that vary with age. Appendages: superiors long, their tips turned downwards over the inferiors. Tips rounded, with small subapical spine. Inferiors comparatively wide, closely apposed, apically upturned. In side view, they are distinctly and deeply indented at about 1/3 of their length. A short spine is found at the level of the indentation. Accessory genitalia: lam. ant. slightly invaginated; ham. ant. forked, but less deeply than in *O. lefebvrei*; ham. post. with an apical, upturned hook. Vesica spermalis with sperm reservoir strongly raised, and with apical shallow, U-shaped invagination to hold the glans. The latter long and narrow, with short flagella.

Female

Head, thorax, wings, legs coloured as in the male. No trace of a depression on the frons in front of the anterior ocellus. Abdomen cyclindrical, marked with black on S_{2-7} as in the male; ultimate segments with diffuse dorsal dark markings. Vulvar scales short, rounded, shallowly excavated; their base not swollen. Floor of S_9 with two arched, sclerotized ridges that touch in a medial point, the inner one not protruding.

Measurements: *Male.* Total length 42–46; abdomen 33–35. *Female.* Total length 41–46; abdomen 31–34.

Distribution: Anatolia, Iran, Afghanistan, Iraq, Armenian S.S.R., Syria, the Lebanon, northern Israel. The species has not yet been recorded from Jordan, expect Wadi Mujib, and seems to reach the limit of its southward extent in the Dead Sea Valley. Capture dates range between May and July.

Israel (Locality records): Massada (7); Jericho (13).

Also recorded from Wadi Mujib (Naḥal Arnon; 13) in Jordan.

Onychogomphus macrodon Selys, 1887

Figs. 260–263, 266–268, 273–277

Onychogomphus macrodon Selys, 1887:24. Schmidt, 1954a:59; Dumont, 1977b:148.

Type Locality: Beirut and Antakya (but holotype lost).

As stated in Dumont (1977b), Selys' male holotype could not be located in the Brussels Museum, and all females associated by Selys with the type turned out to be *O. lefebvrei*. Selys himself, understandably, states that he found it almost impossible to separate the females convincingly. Clearly, the reason for this was that no true females of *O. macrodon* were available to him. Females were first collected by E. Schmidt (1954a), but not fully described. The description given below is based on a couple from Schmidt's collection, kindly donated by S. Asahina.

Male

Head greenish with faint black markings at the base of the labrum and of the frons. Top of frons medially depressed, with a triangular black marking. Vertex and occiput yellow.

Synthorax with carinal and antehumeral stripes dark brown, on a straw-yellow background. Humeral and other lateral synthoracic sutures black. Legs yellow, with black patches at the top of the femora, and a black line on the tibiae.

Wings: Pt brown. Anal field made up of two cells.

Abdomen ochraceous, marked with black on S_{2-6} as in *O. flexuosus*. S_{7-9} with a median black marking, S_{10} with two rather diffuse markings. Appendages: superiors long, their tips bent ventrally, with a series of subapical external spines. Inferiors shorter, their apex upturned, with a very robust sub-basal thickening on each appendix. Accessory genitalia: lam. ant. rather deeply excavated; ham. ant. forked; ham. post. rather short, with broad base and upright apical tooth. Vesica spermalis: sperm reservoir upright, with two lateral wings and a depression between them to accommodate the short and broad apical segment. Flagella rather short.

Female

Head as in male, with a black stripe at the suture between frons and clypeus. Top of frons depressed in the middle, showing a shallow circular imprint in front of the anterior ocellus. A triangular black spot extends into the depression of the frons. The vertex is a flat crest, produced over the lateral ocelli. Occiput yellow, wide and short. Rear of head without oreillettes. Synthorax green. Black markings more reduced than in the male. The same remark holds true for the black markings on the legs. Abdominal segments 2–8 light yellow, sparsely marked with black. Distal half of S_7, S_8 and S_9 with large mid-dorsal black spot. S_{10} entirely yellow; styli yellow. Vulvar scales with angular margins, rather short, shallowly emarginate. No swellings are found at their base. Floor of S_9 with two arched crests, rather widely separated, but united by an extra longitudinal crest.

Measurements (mm): *Male.* Total length 50–52; abdomen 36–37. *Female.* Total length 50; abdomen 35.

Distribution: The range of this little known species seems to be limited to the valleys of the rivers Orontes, Litani, and Jordan. Capture dates range from March to May.

Israel (Locality records): Sedé Neḥemya (1), Deganya (7), Bitanya (7), 'Ubeidiya (7).

Genus PARAGOMPHUS Cowley, 1934
Entomologist, 67:201
(nomen novum for *Mesogomphus* Förster, 1906, preoccupied in Pisces)

Type Species: *Gomphus cognatus* Rambur, 1842.

Size small to medium, with body green or yellow to ochraceous, marked with brown or black. Abdomen with foliations on S_8 in males only. Wings: triangles entire and almost identical in forewing and hind wings. No anal field in hind wing. Male: appendages long, unbranched, slender, apposed, curved downwards. Inferiors much shorter, broad, curved upwards. Vesica spermalis with sperm reservoir deeply excavated, horse shoe-shaped, hiding the "glans", which has long and curled apical flagella. Female: vulvar scales short, very shallowly excavated medially. Floor of S_9 with a semicircular or angular ridge. Field included by this ridge smooth or longitudinally ribbed.

Distribution: Africa, Asia and most of the Mediterranean basin. Two species are regional. In addition, an Oriental species, *P. lineatus* (Selys), reaches eastern Anatolia, while an endemic of north-eastern Africa, *P. pumilio* (Selys), reaches the Nile Delta.

Key to the Species of Paragomphus
(Figs. 208, 278–297)

1. Large species (over 50 mm total length). Synthorax profusely marked with black, including the lateral sutures.
   Male: ham. ant. fork-shaped; ham. post. tapering apically into a blunt hook, curved anteriad and inwards. App. sup. not constricted; app. inf. gently narrowing towards their apex, not angularly constricted.
   Female: semicircular field on floor of S_9 with lateral and apical thickenings and expansions. **Paragomphus sinaiticus** (Morton)

Smaller species (less than 45 mm). Synthorax diffusely marked with black; lateral sutures either unmarked or with faint brown markings.

Male: ham. ant. forcipate, but with an internal apical process; ham. post. broadened apically, almost hammer-shaped,with elongate black comb. App.sup. constricted slightly distal to the apex of the app. inf.; app. inf. with lateral, angular subapical process.

Female: semicircular field on floor of S_9 simple, without thickenings.

Paragomphus genei (Selys)

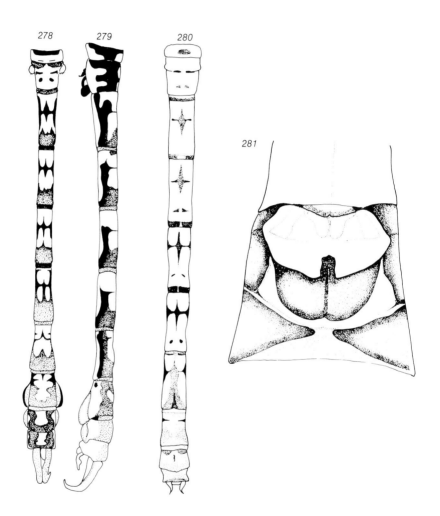

Figs. 278–281: *Paragomphus sinaiticus* (Morton, 1929)
278–279. male abdomen, dorsal and lateral views;
280. female abdomen; 281. valvulae

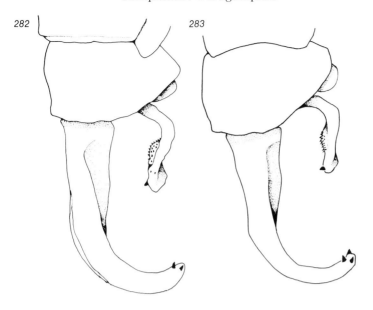

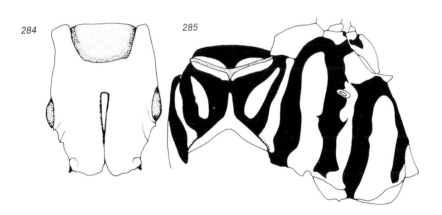

Figs. 282–285: *Paragomphus sinaiticus* (Morton, 1929); male
282–283. appendages in lateral view;
284. appendix inferior; 285. synthorax

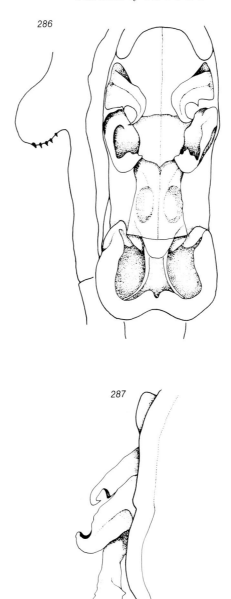

Figs. 286–287: *Paragomphus sinaiticus* (Morton, 1929); male, accessory genitalia, ventral and lateral views

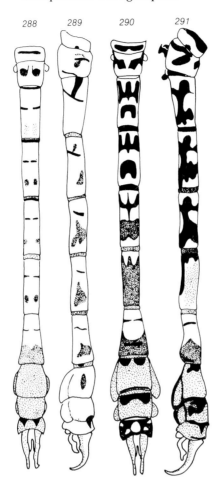

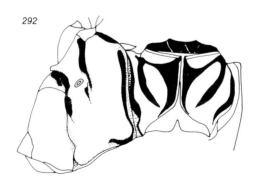

Figs. 288–292: *Paragomphus genei* (Selys, 1841); male
288–291. abdomen, dorsal and lateral views;
288–289. light form; 290–291. dark form;
292. synthorax

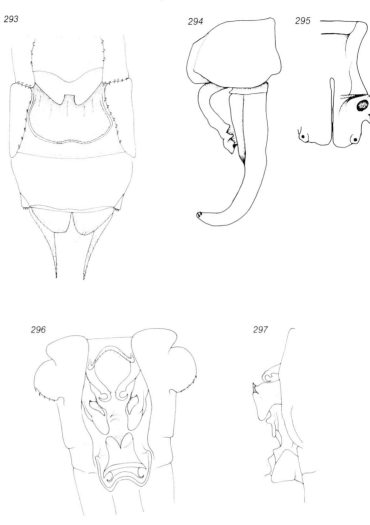

Figs. 293–297: *Paragomphus genei* (Selys, 1841)
293. female, valvulae;
294–297. male;
294. appendages, lateral view; 295. appendix inferior;
296–297. accessory genitalia, ventral and lateral views

Paragomphus sinaiticus (Morton, 1929)

Figs. 278–287

Mesogomphus sinaiticus Morton, 1929:60.
Paragomphus sinaiticus —. St. Quentin, 1965:540; Dumont, 1978c:308.
Paragomphus lineatus (Selys) (in part?). Waterston, 1980:61.

Type Locality: Wadi Feiran, Sinai Mountains.

Male

Mouth parts yellow, labrum, clypeus, frons greyish, with diffuse black markings. Ocellar area black. Vertex pale yellow. Occiput yellow.

Synthorax greenish-yellow, heavily marked with black. Carinal and antehumeral band fused, enclosing an ellipsoidal yellow field in front of synthorax. Humeral, meso-metathoracic and metathoracic sutures covered by three parallel, broad black bands. Legs short, yellow, copiously marked with black.

Pterostigma deep brown. No anal field.

Abdomen ochraceous marked with black; on S_{8-10} these black markings more diffuse. Foliations bright ochraceous. Appendages much longer than S_{10}, robust, not constricted but tapering gently towards their apex. Two to three small black apical spines. Inferiors curved upwards near the base, but with broad flat apical sector set with spinules and with an external subapical spine. In ventral view, the appendix is seen to narrow gently from the level of its inflexion onwards. Accessory genitalia: lam. ant. deeply hollowed-out; ham. ant. simply forked; ham. post. with rather broad base, apically constricted and produced into a blunt, inwardly and anteriorly bent tooth. Vesica spermalis with reservoir deeply excavated; glans long, with long flagella.

Female

Head, thorax, wings and legs as in the male. Abdomen cylindrical, sparsely marked with black. Some rhomboidal mid-dorsal markings on S_{3-6} appear typical of the species. Terminal segments reddish-brown. Styli light yellow, their tips divergent. Vulvar scales rather massively built, very slightly indented medially. Semicircular field on floor of S_9 with lateral and posterior thickenings.

Measurements (mm): *Male.* Total length 50–53; abdomen 36–38. *Female.* Total length 51; abdomen 32.

Distribution: Sinai mountains, and Air mountains (Niger). Waterston (1980) erroneously synonymizes this species with *P. lineatus* (Selys) from India, and reports specimens from Saudi Arabia under that name. It seems almost certain that the specimens from the Red Sea hills (Al Hijaz) captured at Jebel Shammar near Kaybar in April really belong under *P. sinaiticus*.

Capture dates range from April till September. A stagnant-water species.

Sinai (Locality records): Wadi Feiran (22), Wadi Isla (22), Wadi Talh (22).

Paragomphus genei (Selys, 1841)

Figs. 288–297

Gomphus genei Selys, 1841:245. Selys & Hagen, 1850:101, 384.
Onychogomphus genei —. Selys & Hagen, 1854:17; Selys & Hagen, 1857:311.
Onychogomphus Hageni Selys, 1871:15.
Onychogomphus hagenii —. Selys, 1887:28
Mesogomphus Hageni —. Ris, 1909:27; Ris, 1913:468; Ris, 1921:344; Morton, 1924:36.
Mesogomphus genei —. Schmidt, 1954a:60; Conci & Nielsen, 1956:138.
Paragomphus (Mesogomphus) genei —. Schmidt, 1938:147.
Paragomphus hageni —. St. Quentin, 1965:540.
Paragomphus genei —. Lieftinck, 1966:39; Dumont, 1977b:148; Waterston, 1980:61.

Type Locality: Sicily (loc. typ. of *P. hageni* is Egypt).

Male

Head entirely pale yellow, or with some diffuse black markings at base of frons and around ocelli.

Synthorax green with diffuse anterior markings (a complete loop between carinal and antehumeral bands is rare). Lateral sutures brownish. Legs yellow, with fine black stripe near the top of the femur, spines black.

Wings as for genus. Pt yellow to brown.

Abdomen green or ochraceous, variously marked with black and/or brown: two extreme forms are shown in Figs. 288–291. Appendages: app. sup., seen laterally, with a constriction slightly behind the tip of the app. inf. Inferiors curved dorsad, with three to four apical spines. In ventral view, the sides constrict angularly slightly before the apex. Accessory genitalia: lam. ant. shallowly excavated; ham. ant. fork-shaped, with lower point of the fork blunt and an inner apical process on the upper branch of the fork; ham. post. widening apically, hammer-shaped, with a strong black apical crest produced into a forwardly pointing spine. Vesica spermalis: reservoir deeply hollowed-out; glans broad and swollen, flagella rather short.

Female

Colour pattern as in the male. Abdomen cylindrical. Styli long and pointed, yellow. Vulvar scales triangular, indented medially. Floor of S_9 with a U-shaped crest, of the same width throughout.

Measurements (mm): *Male.* Total length 42–47; abdomen 30–36. *Female.* Total length 42–47; abdomen 30–34.

Distribution: The whole of Africa, including the Maghreb countries, Egypt, Saudi Arabia, the Levant, and reaching the limit of its northern extent in the Orontes Valley near Antakya. Also known from some major Mediterranean islands, such as Sicily and Sardinia.

Israel (Locality records): Haifa (3), Wadi Fari'a (6), Binyamina (8), Buteicha Swamps (7), Deganya (7), Wadi Qilt (12), 'Ein Duyuk near Jericho (13). Capture dates range from March to September.

Family CORDULEGASTERIDAE

Large robust dragonflies, coloured black and yellow in both sexes. Head large, eyes meeting in a point, in females sometimes narrowly separated. Labium apically bifid for about one-third of its length; lateral lobes very large. Frons raised above the level of the vertex and sometimes of the occiput. Occiput small, triangular, but continued on rear of head between the eyes. Legs robust, long. Wings variously shaped. Hind wing with base angulate in male, rounded in female. Membranula present. Nodus at centre of wing or distal to it. arc not broken in Palaearctic representatives. d similar in all wings, traversed. Anal loop present, short, composed of about 10 cells. An anal triangle of about 3–6 cells is found in the male only. Abdomen cylindrical, more tumid at base, and dilated from S_7 onwards. Males with lateral oreillettes. Anal appendages: superiors about as long as S_{10}, tapering towards their apex, divaricate, twisted along their long axis and dorso-ventrally compressed, armed with two strong ventral spines. Inferiors quadrate, with apex shallowly emarginate. Accessory genitalia: lam. ant. inconspicuous; ham. ant. strongly developed; ham. post. smaller, tortuous, sickle-shaped. Vesica with sperm reservoir raised posteriorly, with horseshoe-shaped top.

Females with a large ovipositor, composed of two very long valves projecting from the sternite of S_8, closely apposed to form an apically narrowing furrow. In the basal half of the furrow lie a couple of stylets pertaining to S_9. The ovipositor extends well beyond the tip of the abdomen. Oviposition takes place while the female hovers in a vertical position. The long furrow-shaped "sting" bores holes in sand or mud under shallow water, and eggs are pushed into these holes one by one by the internal stylets. The species breed in running waters.

Distribution of the family: Europe, Asia, North Africa (the Atlas Mountains), and North America

One regional genus.

Genus CORDULEGASTER Leach, 1815

Edinburgh Encyclopaedia, 9:136

Type Species: *Libellula boltonii* Donovan, 1807.
Characters as for the family.
One regional species.

Cordulegaster insignis insignis Schneider, 1845

Figs. 298–301

Cordulegaster insignis Schneider, 1845:114. Selys, 1887:34.
Cordulegaster insignis amasina. Morton, 1915a:284. Fraser, 1929:113; Schmidt, 1954a:81.
Cordulegaster insignis insignis —. Morton, 1915a:284; Fraser, 1929:112; Dumont, 1977b:150.

157

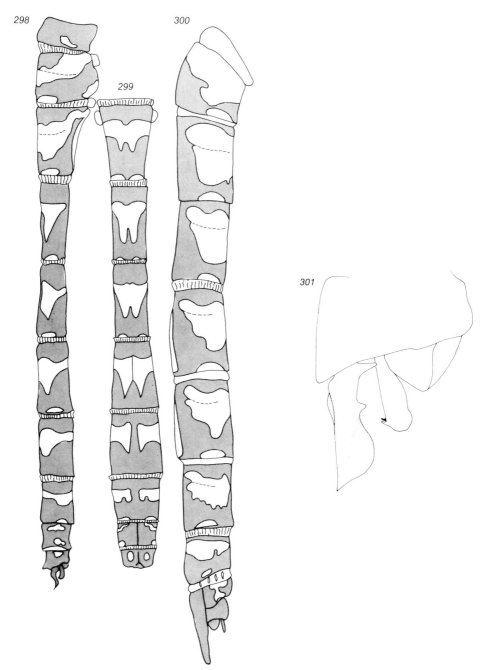

Figs. 298–301: *Cordulegaster insignis insignis* Schneider, 1845; abdomen
298–299. male, dorsal and lateral views; 300. female, lateral view;
301. male terminalia, lateral view

Type Locality: Kellemisch (Gelemish), western Anatolia, Turkey.

Male

Labrum yellow, ventrally emarginate, fringed with brown. Clypeus partly black. Frons yellow, medially emarginate, its base narrowly black. Vertex black. Occiput hairy, tumid, occasionally with black fringe but usually entirely yellow. Rear of occiput yellow.

Synthorax: dorsum black with two oblique cuneiform yellow spots. Sides black with two broad yellow bands, and a third yellow streak, often incomplete, between them. Legs black.

Wings: costa yellow. Pt elongate, brown or black. Anal field not more than 4–5 cells. d entire or with one cross-vein.

Abdomen black, marked with large, medially confluent yellow spots on top of each segment, and two small lunules near the base of the segments. S_9 and S_{10} with small yellow lateral markings only. Appendages: superiors brown-black, parallel, progressively tilted at an angle to one another distally, acutely pointed, their sub-tip deeply indented so that the apices are slightly turned inwards. Ventrally, a strong external basal spine is largely hidden by the apical folds of S_{10}. A second, smaller spine is implanted internally at about 1/4 of the base of the appendix. App. inf. squared, medially indented, with a couple of small upturned hooks on either side. Genitalia as for family.

Female

Coloration as in male, but black margins around labrum wider, and a diffuse transverse black bar on the frons. Abdomen cylindrical, ovipositor entirely black.

Measurements (mm): *Male.* Total length 65–70; abdomen 50–55. *Female.* Total length 63–67; abdomen 49–55.

Distribution: The nominal subspecies occurs in south western and central Anatolia, but probably not in the Pontic area nor in eastern Anatolia , where two vicariant subspecies live. It is also known from Syria and from the Lebanon, where it has definitely been recorded in the Litani Valley. Although the latter river might constitute the southern limit of its range, it is possible that this rivulet-dwelling species lives on streamlets draining Mount Hermon. The species is on the wing from May till September.

Family AESCHNIDAE

Dragonflies of moderate to large and very large size, variable in coloration, but non-metallic. Eyes widely contiguous. Occiput very small, inconspicuous. Labium with middle and lateral lobes about equal in size; middle lobe with a slight median incision at the most. Wings long, base of hind wing in males angulate, or even excavated. d equal in all wings, traversed. Anal loop present. Anal triangle in male usually composed of 3 cells. Membranula well developed. Pterostigma elongate. Legs moderately long, robust. Abdomen in males with oreillettes, constricted at S_9, thereafter cylindrical to its tip. Abdomen swollen at its base in female, gradually tapering thereafter. Anal appendages long but variable in shape. Ovipositor in female complete, often augmented by modifications of ventrum of S_{10} (dentigerous plates), never projecting beyond S_{10}.

Distribution: A large, cosmopolitan family.

Key to the Genera of Aeschnidae
(Figs. 3, 15, 18, 20–22, 23b, 28, 302–333)

1. Anal triangle absent. Hind wing rounded in both sexes 2
–   Anal triangle present. Hind wing angulate at base in male, rounded in female 3
2. S_{4-8} of abdomen with longitudinal supplementary ridges on the sides. Superior appendages of males not pointed. Two rows of cells between Cu and A in hind wing at origin. **Anax** Leach
– S_{4-8} of abdomen without longitudinal supplementary ridges on the sides. Superior appendages of males sharply pointed. Three rows of cells between Cu and A in hind wing at origin. **Hemianax** Selys
3. Cross-veins present in basal space of wing, proximal to arc. Pt short, only slightly longer than wide. **Caliaeschna** Selys
–   No cross-veins in basal space of wing, proximal to arc. Pt long, always several times longer than wide 4
4. Wings saffron yellow, with an amber patch near the base of hind wing. R_3 making an abrupt angle at level of distal corner of pterostigma. Occiput extremely small. **Anaciaeschna** Selys
–   Wings hyaline, no basal coloured spot on hind wing. R_3 not abruptly curved under pterostigma. Occiput readily visible. **Aeshna** Fabricius

Genus ANAX Leach, 1815
Edinburgh Encyclopaedia, 9:137

Type Species: *Anax imperator* Leach, 1815.

Very large and robust dragonflies, variously coloured, non-metallic. Wings hyaline, but often tinted with yellow or brown. Head large. Eyes broadly contiguous. Frons with a crest, not raised. Occiput small. Thorax and legs robust. Wings long and broad, with rounded tips, and rounded base in both sexes. Membranula well developed. Pt long and narrow. d elongate, narrower in forewing than in hind wing, composed of 3–7 cells. ht traversed. Subtriangle absent. Anal triangle absent. Basal space not traversed. Anal loop subquadrate, composed of 3–4 rows of cells, totalling 10–12 in all. Rspl and Mspl deeply curved. R_3 abruptly curved under the pterostigma. Abdomen tumid at base, slightly constricted at S_3, cylindrical or slightly depressed towards its apex. No oreillettes on S_2. Lateral supplementary ridges on S_{4-8}. Anal appendages: superiors much longer than S_{10}, broadly lanceolate, with inner apices rounded, a small external apical spine, and a strong dorsal ridge. Inferior appendages much shorter, broad and quadrate or tapering towards their apex, armed with a few robust spines. Female appendages much shorter than the male's superiors, lanceolate. Ovipositor relatively small. Dentate plate set with fine short spines.

Distribution: Cosmopolitan. Three species occur in Israel and surrounding countries, a fourth one in Saudi Arabia.

Key to the Species of Anax
(Figs. 15, 302–309, 311, 314–315)

1. Synthorax turquoise-blue-green; sides marked with broad black stripes.
                                                                    Anax immaculifrons Rambur
    –   Synthorax grass green, blue-green or pale brownish, without lateral black stripes   2
2. Wings always hyaline. Synthorax grass green or blue-green. Wings: membranula white at base, grey on top. Male with abdomen azure blue, marked with dorsal black. App. sup. completely rounded at their tip. Inferiors about 1/3 the length of superiors, rectangular, without distal spines. Females: abdomen green turning blue-green from S_3 onwards. Occiput without tubercles.   **Anax imperator** Leach
    –   Wings usually at least slightly tinted with yellow or brown. Synthorax brownish, often with light violet sheen. Wings: membranula clear grey. Male: S_2 and base of S_3 azure blue, changing to blue-green from S_3 onwards. The whole dorsum with dark brown markings. App. sup. with external apical spine. App. inf. very short, rounded, with numerous spines along apical margin. Female coloured as male, but darker. Occiput with two tubercles.
                                                                    Anax parthenope (Selys)

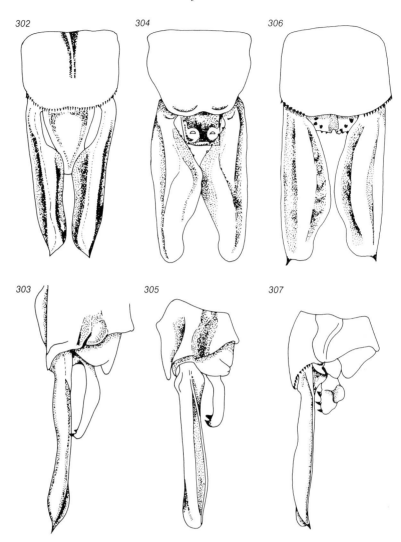

Figs. 302–307: *Anax* spp., male terminalia,
dorsal and lateral views
302–303. *A. immaculifrons* Rambur, 1842;
304–305. *A. imperator* Leach, 1815;
306–307. *A. parthenope* (Selys, 1839)

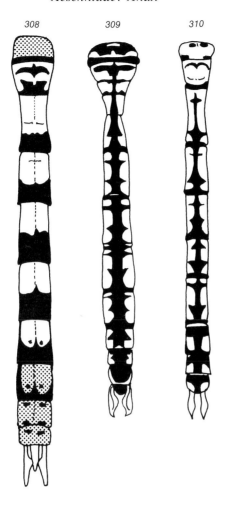

Figs. 308–310: Abdominal markings
308. *Anax immaculifrons* Rambur, 1842; male;
309. *A. imperator* Leach, 1815; male;
310. *Hemianax ephippiger* (Burmeister, 1839); male

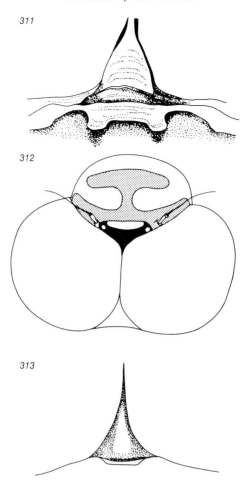

Figs. 311–313: Head structures in various aeschnids
311. Occipital area of *Anax parthenope* (Selys, 1839); female;
312. Head of *Aeshna affinis* Vander Linden, 1820; male;
313. Occiput of *Hemianax ephippiger* (Burmeister, 1839); female

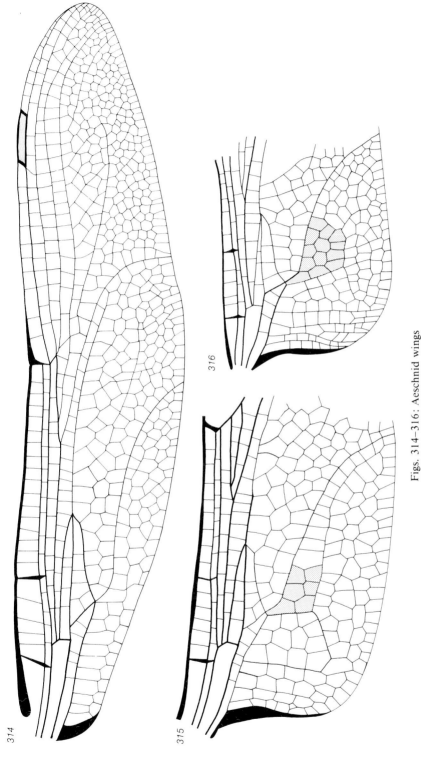

Figs. 314–316: Aeschnid wings

314. *Anax imperator* Leach, 1815; forewing;

315. *A. imperator*, basal half of hind wing (shaded area: two rows of cells between A and Cu);

316. *Hemianax ephippiger* (Burmeister, 1839); basal half of hind wing (shaded area: three rows of cells between A and Cu)

Anax imperator Leach, 1815
Figs. 304–305, 309, 314–315

Anax imperator Leach, 1815:137. Morton, 1924:40; Dumont, 1977b:153; Schneider, 1981a:139.
Aeschna formosa Vander Linden, 1823b:158.
Aeschna (Anax) formosa —. Selys, 1839:387.
Anax formosus —. Rambur, 1842:182; Martin, 1908:9.

Type Locality: "England".

Male
Mouth parts ochraceous. Labrum bright ochraceous, broadly fringed with black. Anteclypeus dark brown. Postclypeus and frons bright greenish-yellow. Top of frons pale yellow, bordered by an azure blue stripe in front, often finely edged with dark brown in front. Base with narrow black border, triangularly produced in the middle. Occiput yellow.
Synthorax grass green with sutures finely brown. Legs black, femora brown on flexor side of apical half to two-thirds.
Wings hyaline, costa yellow. Pterostigma long and narrow, bright ochraceous. Membranula pure white.
Abdomen azure blue, with longitudinal mid-dorsal black markings over its entire length. Anal appendages: superiors blackish-brown, robust, shaped as in Fig. 304. Inferiors less than half the length of superiors, squared, bearing two strong apical spines on each side. Accessory genitalia: lamina medially cleft, with finely pointed anteriorly directed spine; ham. ant. with broad base and U-shaped apex, the opening of the "U" directed inwards, both hamular openings closely apposed; ham. post. small, lanceolate.
Female
Similar to male, but with abdomen shorter and more cylindrical. Venational details of wings as in male. Wings very rarely tinted with pale yellow. Abdomen greenish, with black markings of males replaced by reddish-brown ones. Only sutures between segments black. Anal appendages dark reddish-brown, about 3 times as long as S_{10}, with outer border straight, finely pointed apically. Inner border strongly convex, tapering from base to apex. Styli comparatively short.
Measurements (mm): *Male.* Total length 70–75; abdomen 54–57. *Female.* Total length 66–71; abdomen 50–53.
Distribution: Western and Central Europe, the whole of Africa, Asia Minor and extending eastwards to the north-western provinces of India. Found throughout Israel and Sinai, probably perennial in Sinai.
Israel (Locality records): Dan (1), Buteicha Swamps (7), Deganya (7), 'Ubeidiya (7), Hadera (8), Rosh Ha'Ayin (8), Qishon marshes (5), Tel Aviv (8), Nahal 'Arugot (13), 'En Gedi (13), 'En Avedat (17), Sedé Boqér (17).

Anax parthenope (Selys, 1839)
Figs. 306–307, 311

Aeschna (Anax) parthenope Selys, 1839:389.
Anax parthenope —. Selys, 1840:119; Selys & Hagen, 1850:111; Morton, 1924:40; Fraser, 1936:142; Dumont, 1977b:153; Waterston, 1980:62.

Type Locality: Lake Averno, Naples, Italy.

Male

Head: labium and labrum deep yellow, sometimes with a black border fringing the labrum. Face and frons pale olivaceous, changing to reddish-brown or blackish on top of frons. Base of frons narrowly black. Occiput green-yellow, fringed with black. Legs black, femora reddish-brown.

Wings hyaline, enfumed with yellow or light brown, this tinge becoming darker with age. Pterostigma reddish-brown, long and narrow, overlying three cells. Costa yellow, membranula pale grey, almost white. d in forewing larger and narrower than in hind wing, made up of 4–6 cells in forewing, and usually one cell less in hind wing.

Abdomen: S_1 olivaceous brown, with a sub-basal black spot on each flank. S_2 turquoise blue, with a transverse sub-basal black ridge and a minute black stripe midway. S_{3-10} with a dark brown longitudinal series of markings, as in *Anax imperator*. S_{10} sometimes entirely devoid of black, bluish. Anal appendages: superiors robust, narrowing apically, and with external spine. Inferiors very short, rounded, with about 12 robust spines on each outer corner. Accessory genitalia as in *A. imperator*.

Female

Similar to male, except for crest of frons, which is totally unmarked; blue colour on S_2 more confined to the dorsum of the segment; the sides and the base of S_3 creamy white. Anal appendages dark reddish-brown, as long as the superiors of the male, lanceolate, ribbed. Ovipositor as for genus.

Measurements (mm): *Male.* Total length 64–68; abdomen 47–49. *Female.* Total length 62–67; abdomen 46–48.

Distribution: Southern Europe and North Africa; eastwards reaching Kashmir, southwards reaching India. Occurs throughout Israel and Sinai, mostly on standing water, and almost all year round.

Israel & Sinai (Locality records): Dan (1), Hula (1), Sedé Nehemya (1), Qishon marshes (5), Deganya (7), Rosh Ha'Ayin (8), Jerusalem (11), 'En Gedi (13), Yeroham (17), HaMakhtesh HaGadol (17), Et Tur (23).

Anax immaculifrons Rambur, 1842

Figs. 302–303, 308

Anax immaculifrons Rambur, 1842:189. Martin, 1908:18; Martin, 1909:213; Morton, 1924:40; Gadeau de Kerville, 1926:79; Fraser, 1936:145; Dumont, 1977b:153.

Type Locality: "India".

Male

Labium pale ochraceous, labrum greenish-yellow, broadly fringed with black. Face and frons pale blue-green, with narrow black line at base of frons. Occiput blue.

Synthorax blue-green on dorsum, blue on alar sinuses; carina finely black. Sides turquoise, with broad brown-black humeral stripe covering most of $epst_2$, and an extremely broad band on Su_2 extending anteriorly to level of meso-metathoracic suture. Ventrum black. Legs black.

Wings hyaline, slightly tinted with saffron yellow. Pt brown, rather short (covering not more than 3 cells). Membranula bicolorous, white at base, dark brown at top. d in forewing longer than in hind wing, composed of 5–6 cells in forewing, 4–5 in hind wing.

Abdomen: S_1 entirely brown-black. S_2 turquoise blue with black sutures and a mid-dorsal isolated bat-shaped marking. S_{3-8}: basal half turquoise with pink sheen, distal 1/3 to 1/2 black. S_{9-10} largely brown. Anal appendages: superiors long, pale brown to ochraceous, widened at about 1/3 of their base, apically pointed. Inferiors slightly less than half the length of superiors, tapering towards their apex, which is notched and has one or two small spines on each side. Accessory genitalia as for genus.

Female

Very similar to male, but turquoise blue replaced by pale greenish-blue on synthorax and base of abdomen. Black markings often fringed by reddish-blue, and segment 1 reddish instead of brown. Anal appendages blackish-brown, short. Ovipositor short, as for genus.

Measurements (mm): *Male.* Total length 80–85; abdomen 55–60. *Female.* Total length 81–86; abdomen 58–60.

Distribution: India, Sri Lanka, Pakistan. Not recorded from Iran, but recently found in Anatolia (Dumont, 1977b), and also on the isle of Rhodes (Fisher, pers. comm.). First reported from the Levant by Martin (1909, 1926) at Beit-Meri near Beirut. Morton (1924) found another specimen near Beirut.

Israel (Locality records): Two records, both males, are available: Wadi Qurein, (1; 3.V.1955), and Qusbiya, 10 km S. of Khushniya (18; 3.VI.1973).

A riverine species that breeds only rarely in stagnant water.

Genus HEMIANAX Selys, 1883

Bull. Acad. r. Belg., Ser. 3, 5:723

(*Cyrtosoma* Selys, 1871, preoccupied in Coleoptera)

Type Species: *Aeschna ephippigera* Burmeister, 1839.

Dragonflies of large size, coloured brown, yellow and blue. Eyes broadly contiguous, occiput very small. Frons angled but not crested. Wings of moderate length, their base rounded in both sexes. Membranula well developed. Pt elongate, narrow. d elongate and much longer in forewing than in hind wing. No subtriangle. No anal angle. Anal loop quadrangular, composed of 3–4 rows of cells, about 15 in number. R_3 sharply curved under Pt; Rspl and Mspl deeply concave, even angulate. Basal space not traversed. arc angulate. No oreillettes on S_2. No lateral supplementary ridges on S_{4-8}. Accessory genitalia of male and ovipositor of female as in *Anax*. One species.

Hemianax ephippiger (Burmeister, 1839)

Figs. 310, 313, 316, 321–322

Aeschna ephippigera Burmeister, 1839:840.

Aeschna mediterranea Selys, 1839:391.

Anax mediterranea —. Selys, 1840:120.

Anax senegalensis Rambur, 1842:190.

Anax mediterranaeus —. Selys & Hagen, 1850:329; Hagen, 1863:198.

Hemianax ephippigerus —. Selys, 1883:723; Selys, 1887:36.

Hemianax ephippiger —. Kirby, 1890:85; Martin, 1908:28; Morton, 1924:41; Schmidt, 1938:147; Dumont, 1977b:153; Waterston, 1980:62.

Type Locality: Madras, India.

Male

Mouth parts yellow; labrum with black fringe. Frons yellow. Frontal crest with black base. Base of frons lined with black, thickened medially. Occiput yellow.

Synthorax palest brown or olivaceous, sutures not marked with black except on infra-episterna. Legs black, base of femora brown and inner side of femora of first pair clear yellow.

Wings hyaline, or partly enfumed with pale amber. Hind wing with basal amber patch. Pt golden, long and narrow. Membranula white on top and along its free borders, dark grey elsewhere.

Abdomen bright ochraceous, marked as follows: S_1 and sides of S_2 pale greenish-yellow. Dorsum of S_2 largely azure blue, with narrow black sutures. Some blue still on dorsum of S_3. Most of S_3 and S_{4-7} bright brownish with longitudinal mid-dorsal black stripe, as in *Anax*, but narrower. In addition, a small black spot is found on the sides of S_{3-7}, lengthening to a stripe from S_8 onwards, and becoming confluent with mid-dorsal black band. Clear parts of S_{8-10} yellow, especially S_{10}; Anal

appendages: superiors dagger-shaped, finely pointed, with a mid-dorsal rib and strong subapical hump. Inferiors somewhat less than half the length of superiors, triangular, upper surface covered with strong, imbricate spines. Accessory genitalia: spine on lam. ant. rather strong; base of ham. ant. very broadly built; otherwise, as in *Anax*.
Female
Coloured almost exactly as male, but azure blue on dorsum of S_2 more restricted and ground colour of abdomen deeper brown. Appendages as long as in the male, deep brown with darkened margins, lanceolate. Ovipositor small, as in *Anax*.
Measurements (mm): Male. Total length 65–70; abdomen 52–56. *Female.* Total length 63–69; abdomen 47–50.
Distribution: S. Europe, occasionally reaching W. and C. Europe, and extending to India in the east, to the whole of Africa in the south. A notorious migrant that is often found in large swarms in absolute desert country. Mass migrations in Egypt were recorded by Williams (1925, 1926); the species breeds in shallow, often temporary waters, and is thus preadapted to life in arid environments. Larval development may be as short as 90 days (Gambles, 1960). Found throughout Israel and Sinai almost all year round, except during the coldest months.
Israel & Sinai (Locality records): "Beth Gordon", Deganya A (7), Ḥadera (8), Be'ér Sheva' (15), Jerusalem (11), Wadi Masri near Elat (14), Elat (14), Naḥal Ẕin (17), Isma'iliya (20).

Genus ANACIAESCHNA Selys, 1878
Mitt. k. zool. Mus. Dresden, 3:317

Type Species: *Aeschna jaspidea* Burmeister, 1839.
Dragonflies of large size, variously coloured, with wings always more or less deeply tinted with yellow. Eyes broadly contiguous, and occiput very small. Wings long, rather pointed at apices, base of hind wings angulate in the male. Pt long and narrow. Basal space entire. d in forewing slightly longer than that in hind wing. Subtriangle defined, traversed by a single cross-vein. Hypertrigone traversed. Anal triangle present. IR_3 forked at level of proximal end of Pt. Oreillettes present on S_2 in males. Superior anal appendages long and slender, convex. Females with rather short ovipositor.
Distribution: Europe, Asia, N. America, Africa.
One regional species

Anaciaeschna isoceles antehumeralis (Schmidt, 1950)

Figs. 317–318, 325–326

Libellula quadrifasciata var. β *isoceles* Müller, 1767:125.
Aeschna rufescens Vander Linden, 1825:27. Selys, 1883:729; Selys, 1887:37.
Aeschna isosceles antehumeralis Schmidt, 1950b:8.
Anaciaeschna isoceles humeralis (sic!) —. St. Quentin, 1965:542.
Anaciaeschna isoceles antehumeralis —. St. Quentin, 1964a:424, Dumont, 1977b:152.

Type Locality: Tschiflik, Anatolia.

Male

Mouth parts, face entirely ochraceous. Crest of frons finely bordered with brown-black. Occiput yellow.

Synthorax transparent brown; narrow citron yellow antehumerals, slightly widened apically. Humeral suture and Su_2 narrowly black, widened on infra-episternum. A broad yellow streak on mesepisternum 2; metepisternum 3 entirely yellow. Legs: femora chocolate brown, tibiae black.

Wings: Pt light ochraceous. Membranula large, brownish, slightly clearer at its base. Anal triangle 3–4 celled, covered by a bright amber patch. Other venational characters as for genus.

Abdomen: ground colour brown, with olivaceous sheen on top third of segments. S_1 and S_2 with mid-dorsal bright yellow stripe; sides of both segments including oreillettes greenish-yellow. Abdomen constricted at S_3. Dorsum of all segments with fine tranverse black stripe, black end-rings, and 2–4 diffuse brown spots. A mid-dorsal black stripe on S_8 and top of S_9. Sides of S_{4-9} more greenish, with a diffuse dark brown marking. Appendages: superiors brown, with dark interior margin and tip, elongate, apically pointed, the apex tumid outwardly, and a keel on the apical third. Inferiors triangular, just under half the length of superiors. Accessory genitalia: lam. ant. with strong, posteriorly directed spine, reaching as far as ham. ant., with tip forceps-shaped and an anterior spine. Modified margin of tergite 2 narrow.

Female

Colours as in male. Antehumerals usually not widened. Appendages brown, shorter than superiors of male. Ovipositor rather short but robust.

Measurements (mm): *Male.* Total length 67–70; abdomen 50–52. *Female.* Total length 67–72; abdomen 50–54.

Distribution: The nominal subspecies is found in North Africa, the Iberian Peninsula, Western and Central Europe; ssp. *antehumeralis* occurs in Greece and Anatolia, and reaches the Jordan Valley.

Israel (Locality records): Lake H̱ula (1), Menara (Metulla) (1), Gonén (1), Wadi Qurein (1), Dan (1), Kefar H̱ittim (2), Kefar Masaryk (4).

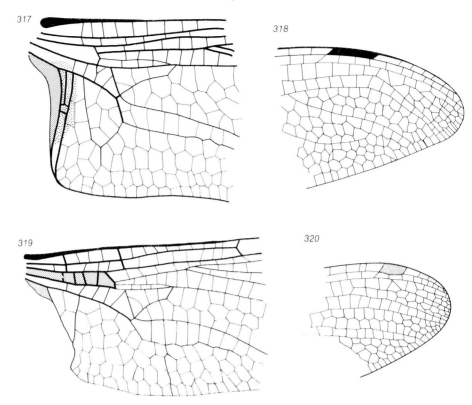

Figs. 317–320: Aeschnid wings
317. *Anaciaeschna isoceles antehumeralis* (Schmidt, 1839), male,
base of hind wing; 318. same species, wing tip
319. *Caliaeschna microstigma* (Schneider, 1845), male,
base of hind wing (shaded area: basal space traversed by cross–veins);
320. same species, wing tip

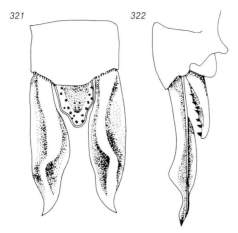

Figs. 321–322: *Hemianax ephippiger* (Burmeister, 1839), male,
terminalia, dorsal and lateral views

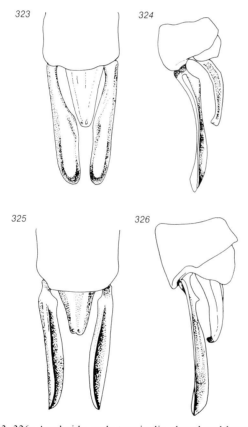

Figs. 323–326: Aeschnids, male terminalia, dorsal and lateral views
323–324. *Caliaeschna microstigma* (Schneider, 1845);
325–326. *Anaciaeschna isoceles antehumeralis* (Schmidt, 1839)

173

Genus AESHNA Fabricius, 1775

Systema Entomologiae, p. 424

Note

The origin of the term *Aeshna* Fabricius is unknown; possible derivations that have been proposed are all speculative. Therefore, no emendation is possible, and the widely used *Aeschna* is invalid. However, the letter c should be preserved in combinations such as *Anaciaeschna* and *Caliaeschna*, which are posterior to (and followed) the erroneous spelling *Aeschna* but are valid in themselves.

Type Species: *Libellula grandis* Linnaeus, 1758.

Dragonflies of large size, often multi-coloured in non-metallic mosaics of green, blue, brown, yellow and black. Wings hyaline, rarely tinted. Eyes very broadly contiguous. Frons angulate, occasionally raised. Occiput very small. Legs of moderate length, robust. Wings long and pointed, base of hind wing broad and angulate in the male, rounded in the female. Membranula well developed. d narrow and elongate, of similar shape in both pairs of wings. Anal triangle present, composed of 3–5 cells. Anal loop variably shaped, composed of 5–12 cells. Basal space entire. Pt very long. Hypertrigones traversed. Oreillettes present in males. Anal appendages: superiors variable in shape. Inferiors simple and triangular. Female with short, lanceolate appendages and a large, robust ovipositor.

Distribution: A large genus, with many species all over the world. Five species occur in Anatolia; one, possibly two, reach the Jordan Valley.

Key to the Species of Aeshna
(Figs. 18, 20–22, 28, 312, 327–332)

1. Sides of synthorax clear blue, turning green ventrally (males) or green all over (females), with all sutures rather broadly marked with black. Male: app. inf. about half the length of app. sup.; app. sup. with sub-basal ventral hump; lam. ant. with very long spine, reaching beyond ham. ant. **Aeshna affinis** Vander Linden

— Sides of synthorax brown, with two yellow streaks; sutures narrowly marked with brown-black. Male: app. inf. well over half the length of app. sup.; app. sup. without sub-basal ventral hump; lam. ant. with short spine, barely reaching ham. ant. **Aeshna mixta** Latreille

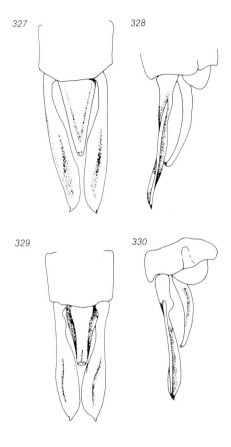

Figs. 327–330: *Aeshna* spp., male,
terminalia, dorsal and lateral views
327–328. *A. mixta* Latreille, 1805;
329–330. *A. affinis* Vander Linden, 1820

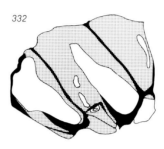

Figs. 331–332: *Aeshna* spp., lateral view of synthorax
331. *A. affinis* Vander Linden, 1820;
332. *A. mixta* Latreille, 1805

Aeshna mixta Latreille, 1805
Figs. 327–328, 332

Aeshna mixta Latreille, 1805:7.
Aeschna mixta —. Morton, 1924:40. Fraser, 1936:130; Dumont, 1977b:152.

Type Locality: Environs of Paris, France.

Male
Mouth parts ochraceous, labrum with black on apex and fine black base. Clypeus and frons light olivaceous. Frons marked with a black "T" on top. Vertex and occiput yellow.
Synthorax transparent, brownish. All sutures narrowly black. Antehumeral band narrow, greenish-yellow, often reduced to a basal spot. Sides with two broad greenish-yellow bands, one on mesepimerum and one on metepimerum, the former tending to narrow, the latter expanding above. Small additional yellow spots may be present between these bands. Legs black, base of femora clear brown.

Wings hyaline. Pt brown. Membranula white, light brownish at free margins. Anal triangle narrow, 3-celled. d of equal size and shape in all wings, 3–4 - celled. IR_3 forked well before proximal edge of Pt, especially in hind wing.

Abdomen with ground colour reddish-brown or black, marked with a mosaic of blue and green spots. Oreillettes small, triangular. Appendages: superiors brown or black, 2 1/2 times as long as S_{10}, narrow at base, then lanceolate, ending in a fine point. Inferiors narrowly triangular, with apex slightly curled up, about 2/3 the length of the superiors. Genitalia: lam. ant. moderately cleft, with a pair of strong spines, turned posteriorly, their tip not reaching the ham. ant.; ham. ant. small, forceps-shaped, without anterior spine; ham. post. lanceolate. Vesica spermalis with sperm reservoir hollowed-out anteriorly, accommodating broad glans. Margins of tergite of S_2 sinuous but little differentiated.

Female

Similar to male, except for the abdomen, which is tumid and not constricted at S_3, and the green-yellow spots which replace the blue-coloured ones in the male. Anal appendages lanceolate, as long as the superiors of the male. Ventrum of S_{10} set with numerous minute spines. Ovipositor well developed.

Measurements : *Male.* Total length 60–64; abdomen 44–47. *Female.* Total length 60–65; abdomen 45–48.

Distribution: Europe, North Africa, Asia Minor and Central Asia, reaching Kashmir in the east.

Israel & Sinai (Locality records): Qishon marshes (5), Gesher (7), Suez Canal (20).

Aeshna affinis Vander Linden, 1820

Figs. 312, 329–331

Aeshna affinis Vander Linden, 1820a:7. Dumont, 1977b:152.
Aeschna affinis —. Selys, 1887:37; Martin, 1908:42.

Type Locality: Bologna, Italy.

Male

Mouth parts and face green or blue. Labrum with fine black base and virgule. Suture between frons and clypeus narrow and black. Top of frons marked with black "T". Vertex and occiput green or blue.

Synthorax transparent, green or blue, brownish in front, with short antehumeral stripes. Sutures broadly marked with black, especially the lower half of the meso-metathoracic suture. No yellow streaks. Legs black, base of femora brown.

Wings: Pt dark brown. d in forewing slightly longer than that in hind wing, 3–4-celled. Anal triangle narrow, 3-celled. Membranula white. IR_3 forked at level of proximal edge of Pt or slightly basal to it.

Abdomen with ground colour brown, extensively marked with blue. Appendages: superiors black, narrow at base, widened and lanceolate, apically pointed. A ventral hump present at the level where the appendix widens. Inferior appendix narrow, triangular, about half the length of superiors. Accessory genitalia: lamina deeply cleft, long and finely pointed, with posteriorly directed spine that extends over the ham. ant.; ham. ant. as for genus, anteriorly pointed; ham. post. lancet-shaped. Sperm reservoir hollowed-out anteriorly. Glans triangular. Oreillettes small. Margins of S_2 not modified.

Female

As male, but blue colours replaced by green and yellow. Anal appendages relatively short, shorter than superiors in male. Ovipositor fairly robust.

Measurements : *Male.* Total length 58–63; abdomen 44–48. *Female.* Total length 58–64; abdomen 45–49.

Distribution: The Mediterranean basin, and extending eastwards as far as Turkestan. The citation from Israel in Dumont (1977b) is erroneous. No definite record from the Levant is as yet available, although the species is likely to occur here.

Genus CALIAESCHNA Selys, 1883
Bull. Acad. r. Belg., Ser. 3, 5:30

Type Species: *Aeschna microstigma* Schneider, 1845.

Moderately large species, with wing tips rounded. Tornus of hind wing angulate in male, rounded in females. Pt short, covering only two subjacent cells, about twice as long as wide. Anal triangle present, wide. Subtriangle present. Hypertrigone traversed. Basal space traversed, mostly with 4 cross-veins. Membranula short and narrow. Eyes contiguous over a short distance only . Triangle larger than in *Aeshna*. Frons raised. Male: oreillettes large. Accessory genitalia: margins of tergite of S_2 modified. Female: ovipositor massive, strongly developed; floor of S_8 raised. Styli not longer than S_{10}. The genus is monotypic, but close to the Oriental *Cephalaeschna*.

Caliaeschna microstigma (Schneider, 1845)

Figs. 319–320, 323–324, 333

Aeschna microstigma Schneider, 1845:113.
Caliaeschna microstigma —. Selys 1883:739; Selys, 1887:37; Morton, 1924:40; Dumont, 1977b:151.

Male

Mouth parts clear yellow; labrum broadly fringed with black. Clypeus and frons light blue, frons raised and produced into a median point, with a fairly wide anterior black patch. Occiput blue.

178

Synthorax brown, with dark blue, comma-shaped antehumerals, and two broad green or light blue lateral bands; sometimes a narrow vertical blue stripe between these oblique bands, and a triangular blue spot between the wings. Legs black, femora brown.

Wings hyaline, venation as for genus.

Abdomen narrow, strongly constricted at S_2. Oreillettes very large. Abdomen deep brown, marked with blue. Anal appendages: superiors long, apically rounded, constricted at their base, with a ventral sub-basal hump. Inferiors triangular, their tip curled upwards, slightly exceeding half the length of the superiors. Accessory genitalia: margin of tergite of S_2 forms a posteriorly produced plate; lam. ant. with a vertical, upright blunt spine on each side; ham. ant. with massive base and internal foliate process; ham. post. short, rounded. Vesica spermalis with reservoir produced into a point between and behind the top of the lamellae of tergite 2. Glans with long, foliate flanges.

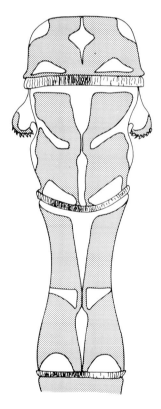

Fig. 333. *Caliaeschna microstigma* (Schneider, 1845), male,
markings on abdominal segments 1–3, dorsal view

Female

Colours as in male, but spot on front of frons sometimes light brown not black. Abdomen not constricted at S_2, tumid, with green-yellow dots instead of blue ones. Styli black, as long as S_{10}. Ovipositor massive, reaching tip of S_{10}. Floor of S_8 triangularly raised.

Measurements (mm): *Male.* Total length 52–59; abdomen 39–44. *Female.* Total length 57–70; abdomen 41–45.

Distribution: From Montenegro to Afghanistan, including northern Iraq, Syria, the Lebanon, Jordan. A running-water species that occurs in the upper reaches of the Jordan River and its tributaries in Israel, between May and October.

Israel (Locality records): Dan (Tel el Qadi; 1), Ḥulata (1), Banias (1).

Family LIBELLULIDAE

Dragonflies of small to large size; coloration variable, occasionally metallic. Eyes always confluent, vertex well developed. Labium with middle lobe very small, not fissured, and lateral lobes very large. Wings variable in shape and width; base of hind wing always rounded in both sexes. d in forewing elongate along the breadth of the wing, usually traversed, always situated well distal to level of arc. d in hind wing elongate in the long axis of the wing, and usually situated at the level of arc or only slightly distal to it. Membranula present, occasionally reduced. Antenodal veins usually numerous, those of the costal space continuous with those of the subcostal space, the last one incomplete in some genera. Primary antenodals indistinguishable from the others. Anal loop elongate, L-shaped. Abdomen cylindrical, triquetral or depressed. S_2 without oreillettes and segment 10 without a keel. Anal appendages simple and rather generalized throughout the family. Only one pair of hamuli present. Genital lobe well developed. Vulvar scales small, rarely elongated.

Key to the Genera of Libellulidae
(Figs. 25–26, 334–445)

1.	Thorax and abdomen with a dark metallic sheen, which may extend uniformly over the body, or alternate with yellow non-metallic markings. Wings partly coloured or hyaline	2
–	Thorax and abdomen variously coloured, but never metallic. Wings variable	3
2.	Wings hyaline. Hind wings not widened at their base. Large insects, well over 4 cm in length. Body colours yellow and metallic blue-black.	**Zygonyx** Hagen
–	Hind wings widened and with a large dark basal metallic patch. Smaller insects, well under 4 cm in length. Body uniformly metallic black.	**Rhyothemis** Hagen
3.	arc at or distal to an_2. Last an in forewing complete. Medium- to large-sized species. Adults, especially males, often blue pruinescent	4

–   arc between an_1 and an_2. Last an in forewing complete or incomplete. Size and colour variable   6

4.   Hind lobe of pronotum small. Base of hind wing with dark brown spot.

                Libellula Linnaeus

–   Hind lobe of pronotum large, erect, fringed with long hairs. Base of hind wing without coloured spot   5

5.   Vertex grooved. Clypeus narrower than frons, i.e. face narrowing downwards.

                Orthetrum Newmann

–   Vertex rounded. Clypeus broader than frons, i.e. face widening downwards.

                Nesciothemis Longfield

6.   Last an in forewing complete   7

–   Last an in forewing incomplete   10

7.   Less than 10 an in forewing   8

–   More than 10 an in forewing   **Libellula** Linnaeus

8.   Eye contact very short. Hind lobe of pronotum large, erect, fringed with long hairs. Small insects with abdomen swollen and dilated in its front half (S_{1-6}), cylindrical and slender in its rear half (S_{7-10}).   **Acisoma** Rambur

–   Eye contact long. Hind lobe of pronotum small, usually naked. Small to moderately sized insects, with abdomen not noticeably swollen in its front half   9

9.   Wings completely hyaline. Wing venation whitish. Subtrigone in forewing consisting of one cell only. Marginal reticulation of wing consisting of a limited number of relatively large cells. Rspl in both pairs of wings short, not more than 5–6, usually 4 cells in length. Small, dark coloured insects.   **Selysiothemis** Ris

–   Hind wing with a brown basal spot. Wing venation dark. Subtrigone in forewing consisting of a minimum of 3 cells. Marginal reticulation of wings a close reticulum of relatively small cells. Rspl more than 7 cells long. Ochraceous-brown coloured insects.

                Urothemis Brauer

10.   Discoidal field (i.e. cell rows projecting beyond d) in hind wing at least slightly expanded at wing margin   11

–   Discoidal field in hind wing parallel or contracted at wing margin   14

11.   Discoidal field starting with two rows of cells. Hind lobe of pronotum moderately large to very large, fringed with long hairs. Small, dark coloured insects   12

–   Discoidal field starting with three rows of cells. Hind lobe of pronotum small   13

12.   Distal tip of anal loop well exceeding distal tip of d. Abdomen in both sexes strongly swollen in its front half (S_{1-6}), contracted and slender afterwards.   **Acisoma** Rambur

–   Distal tip of anal loop not exceeding distal tip of d. Abdomen cylindrical in females, slightly depressed at its base, somewhat triquetral to the end in males.

                Diplacodes Kirby

13.   Males bright red or ochraceous, females ochraceous with amber spot at base of hind wings, and a reduced spot on the forewings. Frons with a deep sulcus, dividing it into two very oblique horse shoe-shaped halves. Vertex low, rounded. 7–14 an.

                Crocothemis Brauer

–   Males dark coloured, brown-black or sometimes dark red. Females brownish. Wing spot, when at wing base, very large and dark brown or black or traversing wing as a band between N and Pt. Females may have hyaline wings. Frons with a shallow sulcus. Vertex high, narrow. 6–8 an.   **Brachythemis** Brauer

14. Hind lobe of pronotum large, erect, fringed with long hairs. Moderately large to small species, with ground colour red or ochraceous. **Sympetrum** Newman
– Hind lobe of pronotum small 15
15. d in forewing and hind wing on same level. Hind wing not conspicuously broadened at its base. Pt of same length in all wings. **Trithemis** Brauer
– d in forewing 3 or more cells distal to d in hind wing. Hind wing conspicuously broadened at base. Pt in forewing longer than that in hind wing. **Pantala** Hagen

Genus ORTHETRUM Newmann, 1883
Entomologist's mon. Mag., 1:511

Type Species: *Libellula coerulescens* Fabricius, 1798.

Robust dragonflies, of variable size. Frons crested, wider than clypeus. Vertex with a groove. Hind rim of pronotum large, erect, emarginate medially, hairy. Legs of moderate length, claw spines at base or middle of claw. Abdomen variously shaped, often more or less constricted at S_3. Lam. ant. usually erect; hamuli species specific. Vesica spermalis with a flagellum and flanges (alae). Vulvar scales little developed. Females with sides of S_8 more or less foliated. Wings long, hind wing broader than forewing. Nodus situated distal to the middle. d in forewing just beyond that in hind wing, and d in hind wing situated at level of arc. arc at an_2 or beyond it. Sectors of arc arise from a common stem. 10–20 an, the last one complete. R_3 strongly bisinuous. Discoidal field in forewing beginning with 3 rows of cells, expanding to 4 or more at wing margin. Anal loop well developed, closed, its distal angle 3 or more cells beyond distal angle of d. Membranula large. Males and sometimes females at maturity covered by a blue pruinosity (except in *O. sabina*; moreover, in the Oriental region, many males are dark red).

Distribution: Cosmopolitan.

Nine species are regional. From Anatolia, 9 species are also known, but 2 are different from those found in the Levant and in Sinai.

Key to the Species of Orthetrum
(Figs. 334–355)

1. Membranula dark brown or black 2
– Membranula white or pale grey 7
2. At least a trace of basal amber on hind wing 3
– No basal amber on hind wing 4
3. Base of abdomen (S_{1-2}) bulbously swollen, abruptly constricted at S_3. No pruinescence in male; appendages yellow. **Orthetrum sabina** (Drury)
– Base of abdomen expanded but not bulbously swollen, gently constricted over the entire length of S_3. Mature males at least in part blue pruinescent; appendages dark 5
4. Small species, with total length less than 4 cm. **Orthetrum taeniolatum** (Schneider)
– Large species, total length exceeding 5 cm. **Orthetrum trinacria** (Selys)

5.   Total length 4 cm or more   6

–   Total length 3.5 cm or less.   **Orthetrum abbotti** Calvert

6.   Male: hamuli with anterior hook broad, turned outwards. Posterior hook forms a ridge ending apically in a hirsute ridge below and external to the anterior hook.
Female: vulvar opening constricted, with thickened lateral margins.

                                                  Orthetrum chrysostigma (Burmeister)

–   Male: hamuli with anterior hook back-turned, acute at tip. Posterior hook a sinuous ridge, ending in a blunt tubercle behind and opposite to the anterior hook.
Female: vulvar opening wide, with thickened base.

                                              Orthetrum brachiale (P. de Beauvois)

7. R_3 distinctly bisinuous   8

– R_3 almost straight, slightly bisinuous in females only.   **Orthetrum ransonneti** (Brauer)

8.   One row of cells between Rs and Rspl.   **Orthetrum anceps** (Schneider)

–   Two full rows of cells between Rs and Rspl, or at least several cell doublings   9

9.   Females (and non-pruinescent males as well) with thick mid-dorsal black stripe across the abdomen. Males: lam. ant. with long hairs; lobus genitalis rectangular. Hamuli: anterior hook small, outwardly turned, pointed. Posterior hook ending in a hirsute ridge below and external to the anterior hook; cleft between them shallow and narrow.

                                                  Orthetrum taeniolatum (Schneider)

–   Females without a mid-dorsal black stripe on the abdomen; only carina blackened. Males: lam. ant. with short hairs; lob. gen. rounded. Hamuli: anterior hook strong, outwardly turned, pointed. Posterior hook rounded, well posterior to and below anterior hook; cleft between them deep and wide.   **Orthetrum brunneum brunneum** (B. de Fonscolombe)

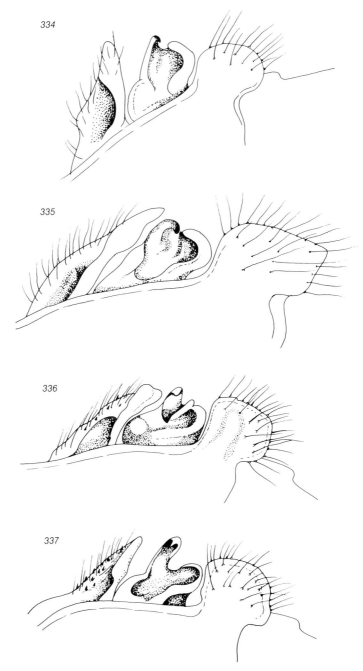

Figs. 334–337: *Orthetrum* spp., males, accessory genitalia, lateral view

334. *O. anceps* (Schneider, 1845);

335. *O. brachiale* (Palisot de Beauvois, 1805);

336. *O. chrysostigma* (Burmeister, 1839);

337. *O. brunneum brunneum* (B. de Fonscolombe, 1837)

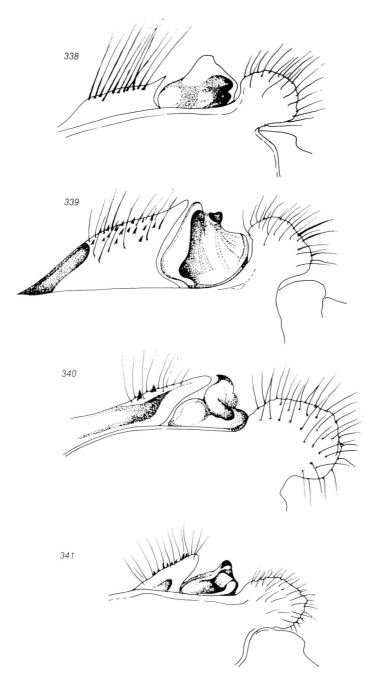

Figs. 338–341. *Orthetrum* spp., males, accessory genitalia, lateral view

338. *O. sabina* (Drury, 1770);

339. *O. trinacria* (Selys, 1841);

340. *O. ransonneti* (Brauer, 1865);

341. *O. taeniolatum* (Schneider, 1845)

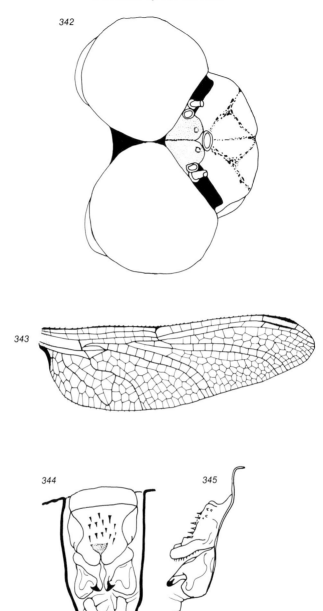

Figs. 342–345: *Orthetrum abbotti* Calvert, 1892; male
342. head, dorsal view; 343. hind wing;
344. accessory genitalia, ventral view; 345. the same, lateral view

186

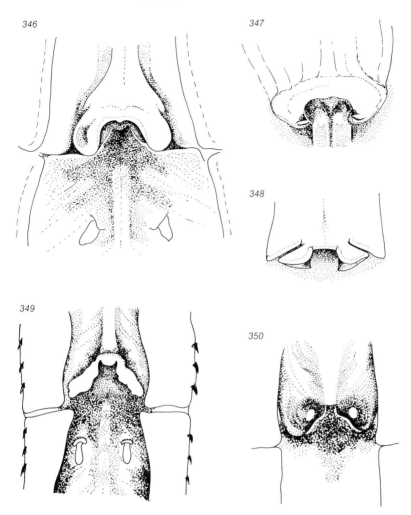

Figs. 346–350: *Orthetrum* spp., vulvar area of females
346. *O. brunneum brunneum* (B. de Fonscolombe, 1837);
347. *O. ransonneti* (Brauer, 1865)
348. *O. chrysostigma* (Burmeister, 1839)
349. *O. taeniolatum* (Schneider, 1845)
350. *O. anceps* (Schneider, 1845)

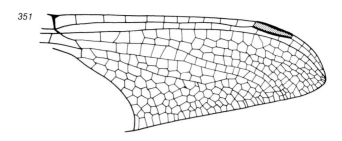

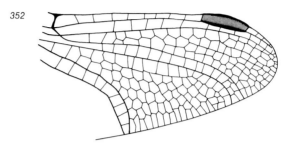

Figs. 351–352. *Orthetrum* spp., apical half of hind wings
351. *O. brunneum brunneum* (B. de Fonscolombe, 1837);
352. *O. anceps* (Schneider, 1845)

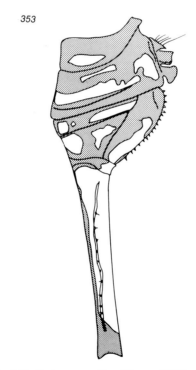

Fig. 353. *Orthetrum sabina* (Drury, 1770), male,
base (S_{1-3}) of abdomen

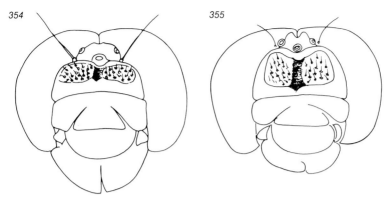

Figs. 354–355: Frontal view of head
354. *Nesciothemis*; 355. *Orthetrum*

Orthetrum brachiale (Palisot de Beauvois, 1805)
Fig. 335

Libellula brachialis Palisot de Beauvois, 1805:171.
Orthetrum brachiale —. Förster (in Kneucker), 1909:44; Longfield, 1955:18; Pinhey, 1970a:
288.

Type Locality: Kingdom of Oware, Benin, southern Nigeria.

Male
Mouth parts greenish, middle lobe of labium and sometimes margins of lateral lobes black. Labrum greenish, with darkened margins and virgule. Clypeus and frons greenish. Frontal shield darkened.
Synthorax green; carinal black stripe very narrow; antehumeral band black, incomplete. Humeral suture narrowly black. Black stripes on inf_2 and on lower half of meso-metathoracic division. Legs black. Basal half of femora with yellow streak.
Wings: Pt large; brown costal edge thick, black. Always two rows of cells in Rspl. Subcostal cross-veins yellow. Membranula dark brown, its base clearer. Always some amber in the hind wing, immmediately bordering the membranula.
Abdomen black, with yellow lateral spots on S_{4-6}; S_{7-9} usually entirely black. S_{10} black or with traces of yellow. Mature specimens with pale blue pruinosity. lam. ant. massive, not bifid at apex, with short hairs. Hamuli broad, with anterior hook turned posteriorly, acute at tip. Posterior hook a sinuous ridge, ending in a rounded tubercle behind the tip of the anterior hook. Genital lobe rounded or rectangular, heavy. Vesica with small sperm reservoir, hardly visible. Apical segment with large flanges and long flagellum.
Female
As male, but not pruinescent at maturity. Vulvar aperture wide, with central crest and lateral swellings. Styli yellow.

189

Measurements (mm): *Male*. Total length 45–50; abdomen 30–33. *Female*. Total length 45–49; abdomen 29–32.

Distribution: Throughout continental Africa south of the Sahara. The species is included in this fauna on evidence of a single citation (Förster, 1909), not confirmed subsequently, from Sinai: Feiran Oasis (22), Wadi Hibran (22) and between Et Tur and Gebel el Hammam (22), March–April. Confusion with *O. chrysostigma* may have occurred, but the presence of the species in the area cannot be *a priori* excluded.

Orthetrum chrysostigma (Burmeister, 1839)

Figs. 336, 348

Libellula chrysostigma Burmeister, 1839:857. Selys, 1887:18.
Libellula barbara Selys, 1849:117.
Orthetrum chrysostigma —. Morton, 1924:41; Morton, 1929:60; Schmidt, 1938:147; Waterston, 1980:63; Schneider, 1981a:140.
Orthetrum chrysostigma chrysostigma —. Dumont, 1977b:158.

Type Locality: Tenerife, Canary Islands.

Male
Mouth parts yellow; face and frons greenish, frontal shield darkened.
Synthorax brownish-green, with narrow black antehumeral, black humeral suture, and an oblique, cream-yellow streak on mesepimerum, bordered by diffuse brown. Su_2 narrow brown. Legs black, femora brown on posterior surface.
Wings: Pt of medium size, brown, costal and radial borders thickly black. Subcostal cross-veins yellow. One or two rows of cells in Rspl. Membranula dark brown. Base of hind wing amber, sometimes reduced to a trace in the basal space only.
Abdomen brown with lateral black stripes on S_{3-9}. At maturity, base of legs, thorax and abdomen completely covered by blue pruinosity. Appendages black: lam. ant. with short hairs and no spines. Apex not bifid. Hamuli: anterior hook broad, erect, turned outwards; posterior hook ending apically in a hirsute ridge below and outward of the anterior hook. Genital lobe angular, hairy.
Female
As male, but less black on synthorax, and legs entirely brown. No lateral black spots on abdomen, except on S_{7-9}. Vulvar opening complex, constricted, angulate at the sides, with sinuous lateral folds.
Measurements (mm): *Male*. Total length 40–48; abdomen 26–33. *Female*. Total length 42–46; abdomen 28–31.
Distribution: Most of Africa, including the Sahara where it is one of the commonest dragonfly species. The larva is pre-adapted to arid conditions in being able to aestivate in damp sand. The species extends through the Arabian Peninsula and the Levant to Iran, Anatolia, Iraq, Afghanistan. It is present in the Oriental region as a distinct

subspecies — *O. chrysostigma luzonicum* (Brauer). In Israel and in Sinai, it is one of the most widespread Anisoptera. In Sinai, it is probably perennial.

Israel & Sinai (Locality records): Sedé Neḥemya (1), Kefar Blum (1), Kefar Ḥittim (2), ʻAkko (4), Nahalal (5), Tiberias (7), Tabigha (7), Bet Sheʼan (7), Migdal (7), Lake Kinneret (7), ʻUbeidiya (7), ʻEn Gév (7), Bitanya (7), Bet Yeraḥ (7), Umm Junni (7), Yarmouk River (7), Ḥadera (8), Hartuv (10), Aqua Bella (11), Wadi Qilt (13), Naḥal ʻArugot (13), ʻEn Gedi (13), ʻEin Duyuk (13), ʻEin el Tureba (13), Naḥal Dawid (13), Wadi Fariʻa (13), Jericho (13), ʻEn Avedat (17), Quseima Oasis (17), Yeroḥam (17), Wadi Hibran (22), Wadi Lubrena (22), Wadi Isla (22), Wadi Feiran (22), Wadi Gharandal (22), Et Tur (23), ʻEin el Furtaga (23), Abu Rudeis (23).

Orthetrum anceps (Schneider, 1845)
Figs. 334, 350, 352

Note

This taxon has alternatively been named *O. anceps* (Schneider) and *O. ramburi* (Selys) in the literature. Hagen (1861), who was the first revisor, decided that one of the types was a female of *O. brunneum*, and therewith dismissed the name. However, W. Schneider (1985d) re-examined *both* type specimens, and found that the second female is the same species as *O. ramburi*, but predates it by three years. The repeated use of the name *anceps* for this taxon, until the 1950s, rules out the possibility of considering it a *nomen oblitum*.

Libellula anceps Schneider, 1845:111.
Libellula Ramburii Selys, 1848:16.
Libellula ramburii —. Hagen, 1863:195; Selys, 1887:14.
Libellula gracilis Selys, 1887:15.
Orthetrum ramburi —. McLachlan, 1889:348; Gadeau de Kerville, 1926:80; Lieftinck, 1966:29; Dumont, 1977b:155.
Orthetrum anceps —. Morton, 1924:41; Morton, 1929:60; Fraser, 1936:295; Schmidt, 1938:147; Schneider, 1985d:97.
Orthetrum coerulescens Fabricius, 1798:285. St. Quentin, 1964c:50; St. Quentin, 1965:543.
Orthetrum coerulescens anceps —. Schmidt, 1954b:252.

Type Locality: Marmaris, Turkey.

Male

Clear brown-ochraceous at emergence, with only end-rings of segments black, soon turning olivaceous, but quickly invaded by a blue pruinosity, covering the whole body. Mouth parts and face of same colour, shield of frons turning dark olivaceous in old specimens.

Legs brown in tenerals, entirely black at maturity.

Wings: Pt of moderate size, brown; subcostal cross-veins yellow. One row of cells in Rspl, rarely one or two cell doublings. At emergence, the wings may be suffused with amber, sometimes as far as the nodus, but this recedes completely within the next day or two. At maturity, no trace of amber is left. Membranula pure white in western Mediterranean and North African populations. In Anatolia and in the Levant, occasional specimens have a grey or light brown membranula. In specimens from Iran and Afghanistan, a brown membranula is the rule. Indian populations have, again, a white membranula.

Accessory genitalia: lam. ant. erect, pointed, bifid apically. Hamuli with anterior hook narrowly pointed, small, above the level of the posterior hook, outwardly pointed. A deep embayment between anterior and posterior hook. The latter a rounded ridge, variable in width with rounded tip. Genital lobe rounded. Appendages brown in juveniles, black in adults. Vesica spermalis with long, rounded flanges and a flagellum.

Female

More robustly built then male. Pt longer. Abdomen brown at maturity, slightly pruinose. Usually a trace of amber at the base of the hind wing. Vulvar aperture: lips strongly swollen, produced medially into a tubercle, so that the vulvar opening appears as a U-shaped invagination between the two lips.

Measurements (mm): *Male.* Total length 36–44; abdomen 23–38. *Female.* Total length 38–45; abdomen 23–29.

Distribution: North Africa, the Levant, Anatolia, the Balkan coasts and the major Mediterranean islands, but not on the Iberian Peninsula (except perhaps the extreme south). Extending east as far as Afghanistan and India. Extremely common in the Levant. Replaced in Europe by *O. coerulescens* (Fabr.) with which introgression appears possible in S. Spain and in the Balkans.

Israel & Sinai (Locality records): 'Ein Jalabina (1), Montfort (1), Naḥal Daliyya at Bat Shelomo (3), Zikhron Ya'aqov (3),'Akko (4), Bet She'an (7), 'Ubeidiya (7), Nabi Rubin (9), Aqua Bella (11), Jericho (13), 'En Gedi (13), 'Ein Fashkha (13), Quseima Oasis (15), 'En Avedat (17), Ramat Magshimim (18), Wasit (18).

Orthetrum abbotti Calvert, 1892

Figs. 342–345

Orthetrum abbotti Calvert, 1892:162. Dumont, 1977d:199.

Type Locality: Mount Kilimanjaro, Kenya.

Male

Mouth parts yellow, labium with central black line. Clypeus and frons greenish. Frontal shield darkened. A thick black line at base of frons.

Synthorax green, with narrow black sutures, soon invaded by azure blue pruinosity. Legs black, base of femora yellow.

Wings: Pt relatively large, dark yellow; subcostal veins yellow; one row of cells in Rspl, occasionally one or two cell doublings. Membranula dark, hind wing with basal amber.

Abdomen: in juveniles S_{4-7} broadly yellow dorsally, black laterally; S_8 black with yellow lateral stripe; S_9 almost completely black. Anal superior appendages straight, black. Accessory genitalia: lam. ant. deeply bifid at apex, with short spines and few hairs. Hamuli: anterior hook long, bent over backwards, finely pointed, with tip turned outwards. Posterior hook a sinuous ridge, rounded posteriorly, much lower than anterior hook. Genital lobe rounded or squarish.

Female

Head like male. Vulvar aperture simple, triangular, lips slightly swollen laterally.

One of the smallest species of *Orthetrum*.

Dimensions (after Longfield, 1955). *Male*. Abdomen 24–25 mm. *Female*. Abdomen 22–25 mm.

Distribution: Most of Africa south of the Sahara.

A single male was found in Wadi Mujib (Naḥal Arnon), Jordan, on 17.VIII.1941 (leg. H. Bytinski-Salz) (Dumont, 1977d). It was exceptionally small; total length 30.5 mm; abdomen 20.5 mm.

Orthetrum taeniolatum (Schneider, 1845)

Figs. 341, 349

Libellula taeniolata Schneider, 1845:111. Hagen, 1863:195; Selys, 1887:17.
Libellula cyprica Hagen, 1863:195.
Orthetrum taeniolatum —. Morton, 1924:41; Andres, 1929:9; Schmidt, 1954a:82; Dumont, 1977a:82.

Type Locality: Kellemisch (Gelemish), western Anatolia, Turkey.

Male

Mouth parts yellow, face darkening towards frons. Base of frons with a black line. Synthorax brown, with creamy antehumeral line and a similar stripe on mesepimerum and metepimerum. Thoracic sutures narrowly black. Legs brown, turning black and pruinescent blue in later life.

Wings: Pt small, rusty, bordered by thick black veins. Subcostal cross-veins yellow. One row of cells in Rspl, but cell doublings frequent, and a complete double row of cells may exceptionally occur. Membranula white in tenerals, tending to become brown in mature specimens. Mature males completely blue pruinescent.

Young males have a light brown abdomen, with characteristic mid-dorsal black stripe on S_{2-10}. Additional black basal streaks are found on S_{4-9}. Accessory genitalia: lam. ant. short, but robust, apically bifid, set with long hairs. Hamuli: anterior hook prominent, triangular, bent slightly backwards. Posterior hook with apical ridge lower and more external than in *O. chrysostigma*, opposite to tip of anterior hook.

Female

Coloured as teneral male, but creamy stripes on synthorax less evident, sometimes not visible at all. Vulvar aperture narrow, U-shaped, with margins thickly swollen. A rather small species.

Measurements (mm): *Male.* Total length 33–38; abdomen 22–26. *Female.* Total length 34–37; abdomen 21–25.

Distribution: Eastern Mediterranean islands, Anatolia, Iran, Iraq, Saudi Arabia and, across the Red Sea, Egypt, Sudan and Somalia. The main range of the species is, however, in India. It is also rather widespread in the Levant.

Israel (Locality records): Migdal (7), Bet Zera' (7), Aqua Bella (11).

Orthetrum ransonneti (Brauer, 1865)

Figs. 340, 347

Libellula ransonneti Brauer, 1865:1009. Selys, 1887:20; Andres, 1929:9; Kimmins, 1934:173; Schmidt, 1954b:253.
Libellula gracilis Selys, 1887:15.

Type Locality: Oasis of Et Tur, Sinai.

Male

Head uniformly pale ochraceous; synthorax and abdomen brown, without markings, except thoracic sutures and end-rings of segments. Legs brown.

Wings: Pt relatively small, ochraceous. Subcostal cross-veins black. One row of cells in Rspl. Membranula narrow, white, very finely bordered with brown. No basal amber.

At maturity, this robust species turns entirely pruinescent blue, and the legs become black.

Accessory genitalia: lam. ant. not bifid at apex, with a few long hairs. Hamulus a triangle, with anterior hook broad at base, and tips turned outwards. Posterior hook no more than a broad basal swelling. Genital lobe large, more or less rectangular, set with long hairs.

Female

Coloured as the male in teneral condition. Vulvar aperture prominent, with deep and wide U-shaped central invagination and strongly swollen sides.

A large and robust species.

Measurements (mm): *Male.* Total length 47–52; abdomen 30–34. *Female.* Total length 45–59; abdomen 28–32.

Distribution: A species typical of arid and hyperarid areas, found in Egypt, Sudan, the Tibesti mountains (Chad Republic), the Hoggar and Air mountains (Algeria and Niger), and the Libyan desert. In Asia it extends from Sinai, probably through Saudi Arabia and eastern Jordan, to eastern Anatolia, Iran, and Afghanistan.

194

Israel & Sinai (Locality records): 'Ein Murra ['En Avedat] (17), 'En Gedi (13), Qadésh Barnéa' (17), HaMakhtesh HaGadol (17), Et Tur (23). Capture dates range from April to September.

Orthetrum brunneum brunneum (B. de Fonscolombe, 1837)

Figs. 337, 346, 351

Libellula brunnea B. de Fonscolombe, 1837:141. Hagen, 1863:195; Selys, 1887:14.

Libellula anceps Schneider, 1845:111 (teste Hagen, 1863).

Orthetrum brunneum —. Ris, 1910:189; Morton, 1924:41; Gadeau de Kerville, 1926:80; Schneider, 1981a:140.

Orthetrum brunneum brunneum —. Schmidt, 1938:147; Dumont, 1977b:156; Waterston, 1980:66.

Type Locality: Provence, France.

Male

Teneral male with head uniformly pale brown. At maturity clypeus and frons grey-blue. Legs brown, black at maturity.

Synthorax clear brown, with short antehumerals, and two clear lateral stripes on the epimeres. Thoracic suture black.

Wings with Pt brown, subcostal cross-veins yellow, and two rows of cells in Rspl. No trace of wing amber. Membranula pure white, rather broad.

Abdomen brown, with black end-rings, a pair of cuneiform black spots in front of the end-rings, a narrow black carinal stripe, and black stripes at the sides of the abdomen. At maturity, thorax and abdomen entirely blue pruinescent. Accessory genitalia: lam. ant. small, very shallowly bifid, bearing very short hairs. Hamuli prominent: anterior hook with an incurvation at about half its length, apex upright and tip turned outwards. Posterior hook a semicircular crest, rounded, lying well below apical hook. Genital lobe rounded, hairy.

Female

As the teneral male; sides of S_8 rather strongly foliate. Vulvar aperture reminiscent of that of *O. ransonneti*.

Measurements (mm): *Male.* Total length 46–49; abdomen 28–32. *Female.* Total length 44–48; abdomen 27–31.

Distribution: Found all around the Mediterranean Sea. North Africa, West and Central Europe, the Balkans. Extremely common in Anatolia and in the Levant. May–August.

Israel & Sinai (Locality records): Tiberias (7), Gesher (7), Hartuv (10), Bethlehem (11), Jerusalem (11), Aqua Bella (11), Jericho (13), Quseima Oasis (17), 'En Avedat (17), Ramat Magshimim (18), Mount Hermon (19), Wadi Talh (22).

Orthetrum sabina (Drury, 1770)

Figs. 338, 353

Libellula sabina Drury, 1770:114. Hagen, 1863:194; Selys, 1887:21.

Lepthemis sabina —. Brauer, 1866:104.

Libellula ampullacea Schneider, 1845:110.

Orthetrum sabina —. Morton, 1924:42; Andres, 1928:30; Schmidt, 1938:148; Dumont, 1977b:158.

Orthetrum sabina ambullacea —. St. Quentin, 1965:543 (*lapsus calami*).

Type Locality: "China".

Male

Mouth parts yellow, face grey, with front of frons black. Legs black, flexor side of femora with long yellow streak.

Synthorax green, with black carina, black antehumerals, humeral suture broadly black and sinuous, and Su_2 equally black. Between these sutures are located two additional oblique black stripes, and a third one behind Su_2. However, there is considerable variation in the degree of development of these stripes, especially the hind one. In very old specimens, the synthorax may turn largely brown, leaving no more than a yellow-green stripe on the mesepimerum.

The wings have a fairly large ochraceous Pt, two rows of cells in Rspl, the apices sometimes narrowly enfumed in old specimens, and a large dark brown membranula with adjacent amber in the hind wing.

The abdomen is green-yellow and black, with typically 4 dorsal and 2 lateral green spots on S_2, lateral spots on S_{3-6}, S_{7-9} entirely black, S_{10} greenish-yellow and appendages whitish-yellow, but all these may become obliterated with age, leaving only the superior appendages yellow. In side view, S_{1-2} are bulbously swollen, constricted distally. S_3 is very narrow and cylindrical; the abdomen widens again at S_{7-9} but much less strongly than at its base. Accessory genitalia: lam. ant. with a very conspicuous tuft of long reddish hair, not bifid at its apex. Hamulus massive, with anterior hook small, surmounted by a ridge that runs anterior to it. Genital lobe rounded, hairy.

Female

Coloured much like the male, but generally paler. Vulvar aperture with lips widened at sides. Styli yellow.

Measurements (mm): Male. Total length 43–50; abdomen 31–36. *Female.* Total length 43–49; abdomen 32–35.

Distribution: North-east Africa, and most of Asia, reaching Australia.

Locally common in Israel and Sinai. May–August.

Israel & Sinai (Locality records): Hula (10), Haifa (3), Bet She'an (7), Rosh Ha'Ayin (8), Sabkhat el Bardawil (20), Et Tur (23), Abu Rudeis (23).

196

Orthetrum trinacria (Selys, 1841)

Fig. 339

Libellula trinacria Selys, 1841:244. Selys & Hagen, 1850:4.
Libellula bremii Rambur, 1842:48.
Lepthemis trinacria —. Selys, 1887:19.
Orthetrum trinacria —. Kirby, 1890:36; Sowerby, 1917:10; Andres, 1928:30; Schmidt, 1938:148; Dumont, 1977b:157.

Type Locality: "Sicily".

Male
Labium and labrum yellow. Frons with black basal line, and a black line in central groove.
Thorax greenish with thin black antehumeral and sutural lines. Legs black.
Wings: Pt long and yellow. Subcostal cross-veins yellow. Rspl with two rows of cells. No basal amber. Membranula dark.
Abdomen long and slender. S_{1-2} in lateral view swollen, constricted towards base of S_2. Unlike in *O. sabina*, the constriction continues on S_3. In juveniles abdomen yellow, copiously marked with black on carina, sides, and end-rings. Anal appendages long, black. At maturity, thorax and abdomen coated with blue pruinosity. Anterior lamina bifid at apex, with spines and hairs. Hamulus rather like in *O. sabina* with anterior hook small, surmounted by a ridge. Genital lobe rounded. Vesica spermalis with slender apical alae.
Female
Much like the immature male. Legs yellow, with black stripes. Styli black, exceptionally long, about twice the length of S_{10}. Foliations inconspicuous, vulvar aperture with lips widened laterally.
A large species, with very long and slender abdomen.
Measurements (mm): *Male.* Total length 54–59; abdomen 40–44. *Female.* Total length 53–58; abdomen 38–42.
Distribution: Most of Africa, including many sites in the Sahara and the Magrheb countries, Egypt, the Levant, and Anatolia. Most of the major Mediterranean islands.
Israel (Locality records): Sedé Neḥemya (1), Rosh Pinna (1), Ḥula (1), Kerach (7), Rosh Ha'Ayin (8).

Genus NESCIOTHEMIS Longfield, 1955

Publções cult. Co. Diam. Angola, 27:59

Type Species: *Orthetrum farinosum* Förster, 1898.
Closely related to *Orthetrum*, from which it differs (Figs. 354–355) in having the vertex rounded; the frons short in depth in front and with two oval depressions; the

clypeus wider than the front of the frons (measured in a vertical direction); the posterior lobe of the pronotum deeper than wide or as deep as wide; the anterior lamina low, hook-shaped and short; the genital lobe very small and inconspicuous; and the vesica spermalis devoid of a flagellum and flanges.

One species is regional.

Nesciothemis farinosum (Förster, 1898)

Fig. 356

Orthetrum farinosum Förster, 1898:169. Andres, 1928:31; Morton, 1929:60.
Nesciothemis farinosum —. Longfield, 1955:59.

Type Locality: Komatipoort, Transvaal, South Africa.

Male

Labium yellow, moderately darkened. Labrum black with yellow margins. Clypeus, frons, and vertex dark, shiny.

Synthorax with carina broadly yellow, humeral suture narrowly black, intermediate space chocolate brown. Sides of synthorax green. Legs greenish-yellow.

Wings: Pt long, ochraceous between thick black nervures; 2 rows of cells in Rspl. Membranula white, with brown margins. No trace of basal amber. Tips of wings slightly enfumed.

Abdomen brown, end-rings of segments black. Some lateral black stripes, especially on the distal half of the abdomen. At maturity, all these colours and markings disappear below a blue pruinescence. Accessory genitalia: lam. ant. backturned; posterior hook large, massive, rounded. Genital lobe very small, elongate. Reservoir of vesica spermalis emerging between the two lobi; apex without flanges or flagellum.

356

Fig. 356: *Nesciothemis farinosum* (Förster, 1898); male,
accessory genitalia, lateral view

Female

Coloured as the teneral male, but Pt even larger, and tips of wings distinctly clouded. Vulvar aperture wide, flat, lips slightly swollen dorsally. Styli black. Foliations on S_8 well developed.

Measurements (mm): *Male.* Total length 42–44; abdomen 22–24. *Female.* Total length 41–43; abdomen 21–24.

Distribution: Widely distributed over continental Africa, and occasionally found in Egypt. No records from Israel or Sinai, but a citation from Suez (Morton, 1929) suggests that the species might occur in the Sinai desert.

Genus LIBELLULA Linnaeus, 1758
Systema Naturae, 10th ed., p. 543

Type Species: *Libellula quadrimaculata* Linnaeus, 1758.

Medium-sized, robust dragonflies, with wings partly coloured. Pronotum with small posterior lobe. Legs short. Wings long, reticulation dense. d in forewing slightly distal to level of d in hind wing, traversed. d in hind wing with base at arc. arc distal or proximal to an_2. Sectors of arc separated from their origin in forewing, shortly fused in hind wing. Distal an complete. Rspl with 2–3 rows of cells. Discoidal field starting with 2–3 rows of cells, markedly widened at wing margin. R_3 bisinuous. Membranula well developed. Pt variable in size. Abdomen triangular or depressed. Males with hamuli small and inconspicuous. Females with small vulvar scales.

Distribution: Europe, Central and Northern Asia, Japan, and North America.

Two species are regional. A third, migratory one (*L. quadrimaculata* L.) is widespread in Europe and North America, and has been found in north Anatolia but not yet in the Levant.

Key to the Species of Libellula
(Figs. 357–360)

1. Forewing with a very fine amber streak between Cu and A, and some amber between Sc and R + M, not exceeding level of arc. Hind wing with brown coloured spot along the membranula, not exceeding level of arc.
    Abdomen rather triangular, not markedly depressed.   **Libellula pontica** (Selys)

–   Forewing with dark brown streak extending between Sc and A from wing base to base of d. Hind wing with broad triangular, brown spot between Sc and wing margin, reaching midway between wing base and N, not including d.
    Abdomen wide and markedly depressed.   **Libellula depressa** (Linnaeus)

Figs. 357–358: *Libellula* spp.
357. *L. pontica* (Selys, 1887), female;
358. *L. depressa* Linnaeus, 1758; male

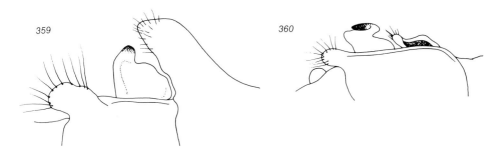

Figs. 359–360: *Libellula* spp., male, accessory genitalia, lateral view
359. *L. depressa* Linnaeus, 1758; 360. *L. pontica* (Selys, 1887)

Libellula pontica (Selys, 1887)

Figs. 357, 360

Libellula fulva race *pontica* Selys, 1887:12.
Libellula conspurcata Schneider, 1845:110.
Libellula fulva Müller, 1764:62. Hagen, 1863:194.
Libellula pontica —. Morton, 1924:42; Schmidt, 1954a:76; Dumont, 1977b:154.

Type Locality: Lake Ḥula, Israel.

Male
Mouth parts bright ochraceous. Clypeus olivaceous. Frons rusty.
Synthorax dull olivaceous, hairy. Legs brown, darkened outwardly.
Wings: Pt black. Main veins reddish, cross-veins yellowish or black. Base of forewings with fine outer streak in subcostal and in cubital area, basal to arc. Hind wing with similar amber in subcostal space, and small basal wing spot: cells filled in with brown, nervures amber. Membranula white. Wing tips hyaline.
Abdomen triangular, reddish-brown, with narrow carinal black. Appendages reddish. Accessory genitalia: lamina depressed; hamuli with strong inner anterior hook, pointed, curved posteriad, and rounded posterior outer hook. Genital lobe small, longish, somewhat twisted and hollowed-out. Adult males may become pruinose blue on abdomen.

Female
Coloured as the male, but carinal black stripe on abdomen wide, especially on S_{7-9}. Vulvar aperture: lips triangularly produced around U-shaped invagination, the bottom of which presents 2 secondary tubercles. Floor of S_9 under the U triangularly produced backwards.
Measurements (mm): *Male*. Total length 39–42; abdomen 23–25. *Female*. Total length 40–41; abdomen 22–23.

Distribution: Anatolia, N. Iraq, Iran, Syria, the Lebanon, and reaching N. Israel. Israel (Locality records): Lake Hula (1), Jabal Jarmak (Har Meron, 1), Meron (1), Sedé Nehemya (1), Dan (1), Benot Ya'aqov bridge on Jordan River (1), Qishon plain (5).

Note

This species is an eastern vicariant of the European *Libellula fulva* Müller, from which it differs in minor characters only, so that it is often considered as a subspecies of the latter.

Libellula depressa Linnaeus, 1758

Figs. 358–359

Libellula depressa Linnaeus, 1758:544. Hagen, 1863:194; Selys, 1887:11; Morton, 1924:42; Schmidt, 1954a:52; Dumont, 1977b:154.

Type Locality: "Southern Sweden".

Male

Mouth parts ochre-coloured, face dark olivaceous, frons dark brown.

Synthorax brown, transparent, hairy. Thoracical sutures black. Legs black, apical 2/3 of femora brown.

Wings: Pt black, main veins black, membranula white. Spots on forewings and hind wings as indicated in the keys.

Abdomen brown, with yellow spots on the sides of S_{3-7} wide and depressed. Tip of abdomen coated with blue pruinosity. Accessory genitalia: lam. ant. large, very deeply cleft, horse shoe-shaped. Hamuli massive, with inner and outer hook. Outer hook rounded, inner hook curved backwards, pointed. Genital lobe rounded.

Female

Differs from the male in that the abdomen is even more depressed and not pruinescent. Dorsum centrally brown, this colour triangularly narrowing from base towards tip, covering the whole breadth of the segment at the base, not more than a carinal stripe at the apex. Sides broadly yellow. Styli as long as S_{10}, brown. Vulvar aperture: lips strongly swollen, vulvar opening broadly U-shaped.

Measurements (mm): *Male.* Total length 45–48; abdomen 25–28. *Female.* Total length 41–43; abdomen 23–25.

Distribution: A spring species, found in Europe, Asia Minor, Syria, the Lebanon. It reaches the limit of its southern extent in Lebanon: Beirut (Morton, 1924). Specimens have, however, been collected from Mount Hermon, Birkat Naqar (19), 22.VI.1971 (Leg. J. Kugler), and from Newé Ativ (19).

Genus SYMPETRUM Newman, 1833

Entomologist's mon. Mag., 1:511

Type Species: *Libellula vulgata* Linnaeus, 1758.

Rather small dragonflies, coloured yellow, brown or reddish with black markings, and wings hyaline or marked with brown and yellow. Pronotum with very strongly developed posterior collar, fringed with long hairs. Legs long and slender. Abdomen cylindrical or triquetral in cross-section, S_8 not dilated in the female. Genitalia variable and species-specific in both sexes. Wings relatively short and broad, reticulation rather open. d in forewing narrow, traversed. d in hind wing situated at base of arc. Sectors of arc shortly fused in forewing, but with longer fusion in hind wing. arc situated between an_1 and an_2. Distal antenodal cross-vein incomplete. 7–9 antenodals. Discoidal field with 3 rows of cells throughout, contracting near wing margin. Rspl with 1 or 2 rows of cells. Membranula moderately large.

Distribution: Europe, Central and Northern Asia, and North America.

Five, possibly six species are regional. No less than 11 species occur in Anatolia.

Key to the Species of Sympetrum
(Figs. 361–388)

1. A broad stria traverses both wing pairs in the area of pterostigma.
   Sympetrum pedemontanum (Allioni)
   – No brown stria across the wings in their apical third   2
2. Legs entirely black.   **Sympetrum sanguineum** (Müller)
   – Legs largely yellow or black with external yellow stripe   3
3. Hind wing with basal amber spot extending across cubito-anal cross-vein and along the membranula. Vulvar aperture of female with lips swollen laterally and vulvar opening in a deep, U-shaped invagination.   **Sympetrum fonscolombei** (Selys)
   – Hind wing with only a trace of basal amber. Vulvar aperture of female not or only very shallowly hollowed-out medially   4
4. Synthorax with black markings extremely reduced: Su_1 and Su_2 barely visible in their lower 2/3, and with a local, elongate black swelling on their upper 1/3.
   Sympetrum meridionale (Selys)
   – Synthorax copiously marked with black   5
5. Both sexes: a black dorso-lateral stripe on S_{2-3}. Male: hamuli with inner hook small, depressed, shorter than outer hook. Ventral angle on app. sup. well defined; tip of app. inf. reaching well beyond it.
   Female: vulvar aperture broad, shallowly excavated medially, but only slightly prominent in side view.   **Sympetrum decoloratum sinaiticum** Dumont
   – Both sexes: no dorso-lateral stripe on S_{2-3}.
   Male: hamuli with inner hook long, upright, longer than outer hook. Ventral angle in app. sup. weakly defined; tip of app. inf. not or only slightly projecting beyond it.
   Female: vulvar aperture very broad, shallowly excavated medially, very prominent in side view.   **Sympetrum striolatum striolatum** (Charpentier)

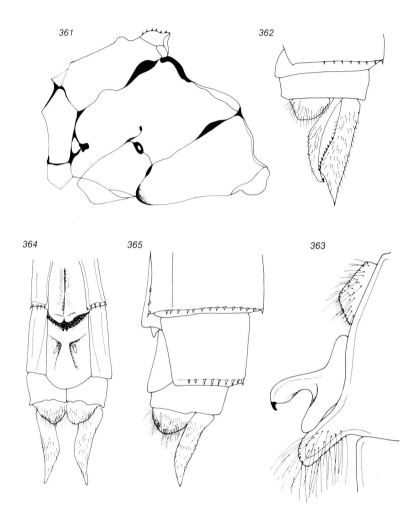

Figs. 361–365: *Sympetrum meridionale* (Selys, 1841)
361. synthorax, lateral view; 362. male appendages, lateral view;
363. male accessory genitalia, lateral view;
364–365. vulvar area and terminalia, female, ventral and lateral views

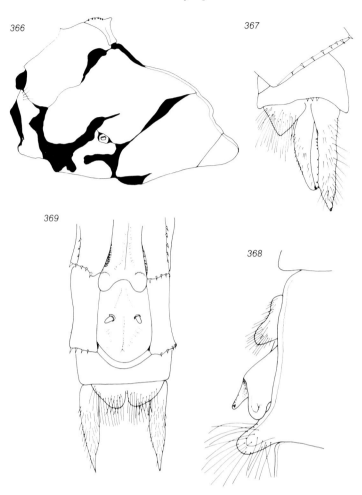

Figs. 366–369: *Sympetrum fonscolombei* (Selys, 1837)
366. synthorax, lateral view; 367. male appendages, lateral view;
368. male accessory genitalia, lateral view;
369. female terminalia and vulvar area, ventral view

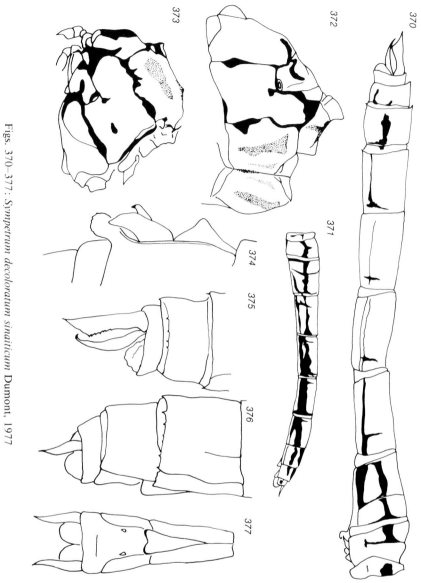

Figs. 370–377: *Sympetrum decoloratum sinaiticum* Dumont, 1977

370. male, abdomen; 371. female, abdomen; 372. male, synthorax; 373. female, synthorax; 374. male accessory genitalia, lateral view; 375. male terminalia, lateral view; 376. female terminalia, lateral view; 377. female terminalia and vulvar area, ventral view

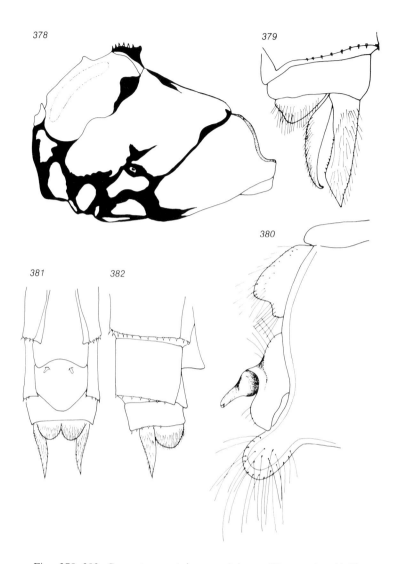

Figs. 378–382: *Sympetrum striolatum striolatum* (Charpentier, 1840)
378. synthorax, male; 379. male terminalia, lateral view;
380. male accessory genitalia, lateral view;
381. female terminalia and vulvar area, ventral view; 382. the same, lateral view

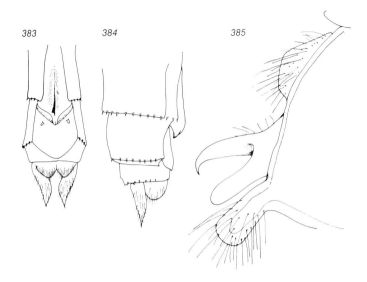

Figs. 383–385: *Sympetrum sanguineum* (Müller, 1764)
383–384. female terminalia and vulvar area, ventral and lateral views;
385. male accessory genitalia, lateral view

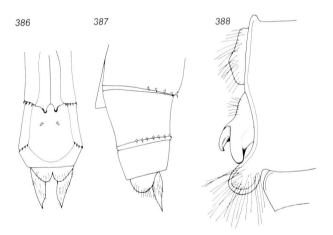

Figs. 386–388: *Sympetrum pedemontanum* (Allioni, 1766)
386–387. female terminalia and vulvar area, ventral and lateral views;
388. male accessory genitalia, lateral view

Sympetrum meridionale (Selys, 1841)
Figs. 361–365

Libellula meridionalis Selys, 1841:245.
Diplax meridionalis —. Brauer, 1868:370; Gadeau de Kerville, 1926:80.
Sympetrum meridionale —. Meyer-Dür, 1874:326; Selys, 1887:11; Morton, 1924:43; Schmidt, 1938:148; Schmidt, 1961:427; Dumont, 1977b:159.

Type Locality: Island of Sardinia, central Mediterranean.

Male
Head: labium yellow, labrum ochraceous, clypeus greenish, frons ochraceous. Base of frons with a black stripe.
Synthorax olivaceous or green-yellow with very reduced black markings. Carina often flanked by a dorsally narrowing stripe on either side; antehumeral stripes brown, rarely absent. Humeral sutures and Su_2 only very narrowly black, with a lense-shaped black spot on their upper third only. Legs bright yellow except for internal (flexor) surface, which is black.
Wings hyaline, with a trace of basal amber. Membranula white or grey, bordered by a conspicuous black nervure. Pt brown.
Abdomen yellow to brilliant red on dorsum, according to age. End-rings black. S_1 with broad black basal stripe. Paired black dots on each segment. Anal appendages red.
Appendages: tips of inferiors just surpassing the level of the ventral angle in the superiors. Accessory genitalia: lam. ant. excavated, low. Hamuli: outer branch long, rounded at tip, inner branch even longer, acutely pointed at tip, upright. Genital lobe elongate, hairy.
Female
Even paler than the male. Ground colour more yellow or ochraceous than red. Basal lateral stripes on S_{3-9} narrow. A narrow carinal black stripe, somewhat widened on S_2 and S_9 only. Styli ochre-coloured.
Measurements (mm): *Male.* Total length 36–40; abdomen 25–28. *Female.* Total length 35–39; abdomen 24–27.
Distribution: South Europe, North Africa and, through Asia Minor, extending as far east as Kashmir. A summer species, relatively common in the Levant. Not found in true desert country. April–September.
Israel (Locality records): Nazareth (2), Qishon marshes (5), Migdal (7), Deganya (7), Ḥadera (8).

Sympetrum fonscolombei (Selys, 1837)

Figs. 366–369

Libellula fonscolombi Selys, 1837:23.
Libellula fonscolombii —. Selys, 1840:49; Hagen, 1863:196.
Libellula erythroneura Schneider, 1845:111.
Diplax fonscolombei —. Brauer, 1866:104; Gadeau de Kerville, 1926:80.
Sympetrum fonscolombei —. Meyer-Dür, 1874:327; Morton, 1924:43; Schmidt, 1938:148; Dumont, 1977b:159; Schmidt, 1961:425; Waterston, 1980:64; Schneider, 1981a:141.
Sympetrum fonscolombii —. Selys, 1887:9.

Type Locality: Aix-en-Provence, France.

Male

Labium yellow, labrum ochre, later turning red. Clypeus greenish, frons reddish-brown. A broad black stripe at base of frons. Vertex and occiput yellow, with reddish sheen. In mature specimens, the whole face turns bright red.

Synthorax greenish, with reddish-brown sheen in mature specimens, coated with long hairs. usually no antehumerals. Sutures marked with black, a long black streak on meso-metathoracic suture. Legs black, external (extensor) surface with bright yellow stripe.

Wings: main veins yellow in tenerals, red in adults. Base of all wings with amber, forming a medium-sized spot on hind wing. Membranula white or pale grey.

Abdomen bright ochraceous at first, turning blood red at maturity. Some black on the dorsum of S_1 and top and sides of S_2. Dorsum of S_{8-9} also with carinal and lateral black stripes. In mature specimens, basal lateral stripes may extend to S_{3-4} and to S_8. In desert environments, however, very pale forms may occur, in which only the dorsum of the segments turns red, the sides remaining pale yellow. Anal appendages long and slim, yellow, later bright red. Accessory genitalia: lam. ant. very low, shallowly excavated, black. Hamuli small, with broad, short outer ramus, and much shorter pointed inner ramus. Genital lobe yellow, elongate, hairy.

Female

Ground colour as that of the juvenile male, rarely turning red. Abdomen more copiously marked with black; a basal black stripe at the sides of the abdomen runs posteriad, gradually narrowing along its course. A second series of black stripes occurs on the flanks, and extends from S_{4-6}, while it is vestigial on S_3. A carinal black spot present on S_8 and S_9, and a fine line on S_{10}. Vulvar aperture reminescent of an *Orthetrum*: only slightly projecting laterally, the lips are swollen, and a deep U-shaped invagination is found in the centre.

Measurements (mm): *Male.* Total length 33–39; abdomen 22–28. *Female.* Total length 34–39; abdomen 23–27.

Distribution: South and East Europe, occasionally reaching Western Europe. The whole of Africa, and extending into Asia as far east as Kashmir. Very widely distributed in the Levant. April–October.

Israel & Sinai (Locality records): Dan (1), Bet Zera' (7), Gesher (7), Rosh Ha'Ayin (8), Ashqelon (9), Jerusalem (11), Nizzana (15), Birkat Ram (18), Wadi Feiran (22), Wadi Tayiba (23).

Sympetrum decoloratum sinaiticum Dumont, 1977
Figs. 370–377

Sympetrum decoloratum Selys, 1884:35. Ris, 1911:629 (pars).

Sympetrum decoloratum —. Le Roi, 1915:614; Ris, 1911:646; Lacroix, 1924:219; Morton, 1924:43.

Sympetrum decoloratum sinaiticum Dumont, 1977a:83.

Type Locality: Oasis of Tozeur, Tunisia.

Male

Face yellow, frons somewhat darkened. Some black at base of vertex.

Synthorax olivaceous, marked with black on sutures as in Fig. 372. Antehumerals diffuse or absent. Stripe on meso-metathoracic suture narrow. Legs yellow, flexor side black.

Wings: venation dark brown, costa lighter. No more than a trace of amber at the base of the wings.

Abdomen yellow, turning reddish-brown with pale sides, marked with lateral black stripes as in Fig. 370. A distinct dorso-lateral stripe on S_{2-3}. No mid-dorsal black markings on S_{8-9}. Accessory genitalia: lamina depressed. Hamuli small, with outer branch broad at base, slightly tapering towards apex, rounded. Inner branch short, curved, pointed. Genital lobe rounded, somewhat elongate. Appendages: superiors bright ochraceous. Tips of inferiors black. Inferiors extending well beyond the ventral angle in the superiors.

Female

Straw-coloured like the teneral male, but antehumerals at least vestigial, and sides of abdomen more copiously marked with a (partially) double row of black stripes (Fig. 371), and mid-dorsal black markings on S_{8-9} present. Vulvar aperture: not prominent laterally, slightly raised, with straight or shallowly concave posterior margin.

Measurements (mm): Male. Total length 34–37; abdomen 23–26. *Female.* Total length 33–36; abdomen 23–25.

Distribution: The nominal subspecies (type locality: Tartum) is found in eastern Anatolia, part of the Caucasus, Iran, Turkestan, Afghanistan and probably also elsewhere in Central Asia. Ssp. *sinaiticum* extends from southern Tunisia and the Algerian Sahara, through Libya and Egypt, to Sinai. In the west, it probably reaches the Iberian Peninsula, although it has not yet been recorded from Morocco.

The nominal subspecies is smaller and both subspecies appear to be separated by a gap across the 'Arava, Jordan, Litani and Orontes Valleys, but possibly the species has been overlooked in intermediate localities. Specimens in the collection of "Beth Gordon" (Kibbutz Deganya A) identified as *S. decoloratum* by Erich Schmidt (Bitanya, 15.V.1931), if confirmed, would support this hypothesis. Flight records range from May to September.

Sinai (Locality records): Wadi el Arbain (22), Gebel Katharina (22).

Sympetrum striolatum striolatum (Charpentier, 1840)

Figs. 378–382

Libellula striolata Charpentier, 1840:78. Schneider, 1845:112; Hagen, 1863:196.

Diplax striolata —. Gadeau de Kerville, 1926:79.

Sympetrum striolatum —. Selys, 1887:10; Ris, 1911:631; Morton, 1924:43; Schmidt, 1961:427; St. Quentin, 1964c:50; St. Quentin, 1965:544.

Sympetrum striolatum striolatum —. Schmidt, 1954b:256; Dumont, 1977b:161.

Type Locality: "Silesia".

Male

Mouth parts yellow; face greenish, totally unmarked. A black bar between frons and vertex (yellow), not descending along the margins of the compound eyes. Occiput brown-yellow.

Synthorax: front brown, with yellow elongated antehumeral stripes in tenerals, later obliterated as the front of the synthorax turns uniformly brown or reddish-brown. Su$_1$ and Su$_2$ broadly black, and intermediate anastomosing black markings on a bright yellow background as in Fig. 378. Legs: tibiae yellow, flexor sides darker than extensor sides. Flexor side of femora black, extensor side with fine yellow stripe.

Wings with base narrowly and diffusely clouded with amber. Membranula white. Pt brown.

Abdomen brown, later bright red, with black dots near the top of each segment. A basal line along the abdominal segments, and lateral stripes on S$_{8-9}$. Appendages ochre-coloured, later turning red. Superiors with ventral angle weakly developed, lined with a row of black spines. Inferiors ochre-coloured as well, with their tip at the level of the ventral angle in the superiors, or only very slightly beyond it. Accessory genitalia: lamina low. Hamuli with outer branch apically tapering but blunt. Inner branch long, upright, finely pointed at its apex, the latter slightly curved backwards. Genital lobe massive, rounded.

Female

Coloured somewhat like the teneral male, but all colours more dull and less contrasting (e.g. the side of the synthorax with olivaceous rather than yellow tinges). Abdomen brownish with a basal lateral black stripe and fine lateral black streaks that widen towards the terminal segments. No dorso-lateral black streaks on S$_{2-3}$. A fine carinal black stripe and fine lateral black streaks that widen towards the terminal

212

segments. No dorso-lateral black streaks on S_{2-3}. A narrow carinal black stripe, widening on S_{8-9}. Vulvar aperture very broad, prominent in side view, with shallow median excavation.

Measurements (mm): *Male*. Total length 35–39; abdomen 21–25. *Female*. Total length 36–41; abdomen 22–27.

Distribution: The Maghreb countries, Europe and, across Anatolia, eastwards as far as Kashmir. In Central Asia, a distinct subspecies (*S.s. pallidum*) is found. Anatolian records refer to the nominal subspecies and are summed up in Dumont (1977b).

Locality records: The species has been recorded from pine-woods at Brummana, the Lebanon (Gadeau de Kerville, 1926) and recently also from Qusbiya (18).

Sympetrum sanguineum (Müller, 1764)
Figs. 383–385

Libellula sanguinea Müller, 1764:62. Schneider, 1845:112.

Diplax sanguinea —. Brauer, 1868:720.

Sympetrum sanguineum —. Meyer-Dür, 1874:328; Selys, 1887:9; Ris, 1911:643; Morton, 1924:43; St. Quentin, 1964c:50; Dumont, 1977b:160.

Type Locality: "Denmark".

Male

Mouth parts and face clear in young specimens, turning red at maturity. Labrum with fine black margin and rudimentary black virgule. A broad black band at the base of the frons.

Synthorax brownish, hairy, reddish at maturity, rather heavily marked with black especially along humeral suture. No antehumerals, but anterior margin of pronotum and synthorax very broadly black. Legs entirely black, except for their extreme base.

Wings: venation black. A well-defined amber spot at the base of the hind wing, reaching the cubital cross-vein. Membranula white or light grey.

Abdomen ochraceous, later turning bright red. S_{8-9} with narrow carinal black markings, sides with black streak from S_3 onwards, but tending to extend on ventrum rather than on sides. Appendages brown, turning red. Inferiors short, their tip at the level of the angle in the superiors. Accessory genitalia: lamina depressed. Hamulus with broad base. External ramus rather long, rounded; internal hook longer, slightly tumid at base, very sharply hooked. Genital lobe rounded, longish.

Female

Coloured like the teneral male; front of synthorax often brown or olivaceous, sides yellow or green. Flanks of abdomen with a single broad basal stripe. Vulvar aperture produced into a slightly bifid point.

Measurements (mm): *Male*. Total length 35–38; abdomen 21–24. *Female*. Total length 36–39; abdomen 22–26.

Distribution: Most of Europe, North and Central Asia, North Africa. In eastern Anatolia and the Caucasus, a distinct subspecies, *S. s. armeniacum* (Selys), is found. May–October.

Israel (Locality record): H̱ula (1).

Sympetrum pedemontanum (Allioni, 1766)
Figs. 386–388

Libellula pedemontana Allioni, 1766:194.
Diplax pedemontana —. Brauer, 1868:720.
Sympetrum pedemontanum —. Meyer-Dür, 1874:328; Selys, 1887:975; Bodenheimer, 1937:231; Dumont, 1977b:161.

Type Locality: Area of Torino, Italy.

Male
Mouth parts yellow, central lobe of labium at times black. Face greenish. Base of frons broadly black.

Front of synthorax brown, sides greenish-yellow. Spot on upper part of humeral suture thickened, on meso-metathoracic suture narrow or absent. Legs black.

Wings with close apical reticulum, and transverse black band, the outer margin of which touches the proximal end of Pt. Some basal wing amber present. Membranula white.

Abdomen brown or red. A mid-dorsal streak on S_9, and a very narrow black lateral stripe along S_{3-9}, somewhat widened on S_{8-9}. Accessory genitalia: lamina very small, hairy. Hamuli rather small, external branch tapering towards rounded apex, inner branch shorter, curved, apically hooked. Genital lobe elongate.

Female
Coloured as the teneral male, but contrast between brown front and yellow sides of synthorax stronger, and base of segments with very broad black band, extending to ventrum of segments. S_{8-9} with carinal black stripe. Vulvar aperture: lip produced laterally; medial opening U-shaped with oblique walls. Floor of S_9 produced backwards, tip rounded.

A small species.

Measurements (mm): Male. Total length 29–35; abdomen 18–23. *Female.* Total length 28–34; abdomen 18–33.

Distribution: Most of Europe, North and Central Asia. Recorded by Selys (1887) from several localities in Anatolia, and by Akramowski (1948) from the Armenian S.S.R. No precise records from the Levant are available, but its occurrence as far south as the Jordan Valley is conceivable. June–September.

Genus CROCOTHEMIS Brauer, 1868

Verh. zool.-bot. Ges. Wien, 18:367

Type Species: *Libellula servilia* Drury, 1770.

Moderately-sized, robust dragonflies. Males coloured uniformly red, females brownish. Pronotum with small posterior lobe. Legs rather short. Abdomen depressed. Wings hyaline or partly coloured at base. Reticulation close. d in forewing narrow, usually traversed; in hind wing with base at arc, usually entire. arc situated between an_1 and an_2. Distal antenodal cross-vein incomplete. 14 an. Discoidal field begins with 3 cell rows, divergent at wing margin. Pt and membranula large. Accessory genitalia of male: lamina depressed; base of hamuli rectangular, with outer branch narrowing distally in side view, foliate in ventral view; inner hook strongly curved, with one to several apical and subapical spines. Genital lobe elongate oval, bent over backwards. Females without foliations on S_8, vulvar valvules long.

Distribution: Africa, Europe, Asia and Australia. Recently also found in the south of the U.S.A. (introduced?).

Two species are regional.

Key to the Species of Crocothemis
(Figs. 389–403)

1. Small species, with abdomen rather narrow (2.5–3 mm in width) and Pt under 3 mm in length. Lateral carina of abdomen with relatively few denticles, e.g. S_5 with 10–12 denticles. Abdomen of male deep red, with at least traces of black lateral lines on some segments. Interior branch of hamulus visible in lateral view. Genital lobe elongate. Female with vulvar scales scoop-shaped, massive, reaching near to the tip of S_{10}.

    Crocothemis sanguinolenta arabica Schneider

–   Large species, with abdomen depressed (over 3 mm in width), and Pt over 3 mm in length. Lateral carina of abdomen with more denticles, e.g. S_5 with 17–22 denticles. Abdomen of male deep red, without any black, except on carina of $S_{8–9}$. Interior branch of hamulus not visible in lateral view. Genital lobe rounded or squarish. Female with vulvar scales scoop-shaped but short, rarely emerging beyond S_9 2

2. Wing apices smoky; costal and subcostal fields suffused with yellow. Male: inner branch of hamulus with a single apical hook. Vesica spermalis with prominent apical vesicle, with upturned apex. Female: valvulae vulvae with a basal swelling, directed apically.

    Crocothemis servilia (Drury)

–   Wing apices, costal and subcostal fields hyaline. Male: inner branch of hamulus with bifid apex. Vesica spermalis with median apical vesicle largely hidden behind protruding lateral walls of basal capsule. Female: valvulae vulvae with weakly developed basal swelling, pointing posteriad.   **Crocothemis erythraea** (Brullé)

389

390

Figs. 389–390: *Crocothemis* spp., males
389. *C. erythraea* (Brullé, 1832);
390. *C. servilia* (Drury, 1770)

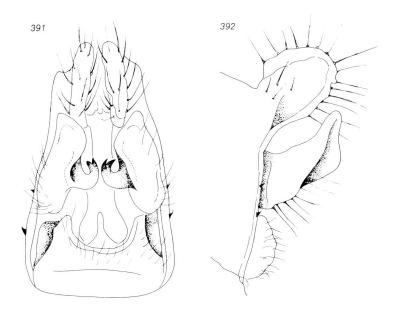

Figs. 391–392: *Crocothemis erythraea* (Brullé, 1832), male,
accessory genitalia, ventral and lateral views

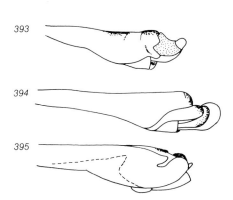

Figs. 393–395: *Crocothemis* spp., males,
vesica spermalis (after Lohman, 1981)
393. *C. servilia* (Drury, 1770);
394. *C. sanguinolenta* (Burmeister, 1839);
395. *C. erythraea* (Brullé, 1832)

217

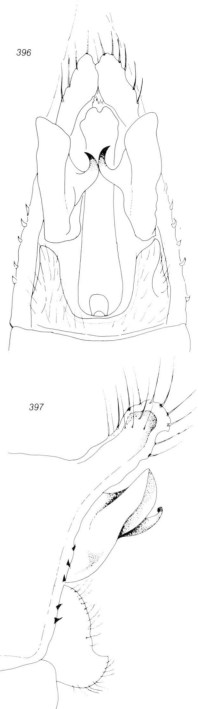

Figs. 396–397: *Crocothemis sanguinolenta arabica* (Schneider, 1982), male, accessory genitalia, ventral and lateral views

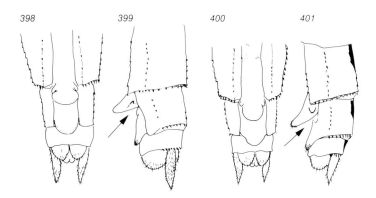

Figs. 398–401: *Crocothemis* spp., female,
terminalia and vulvar area, ventral and lateral views
398–399. *C. servilia* (Drury, 1770);
400–401. *C. erythraea* (Brullé, 1832)
(after Schneider, 1984)

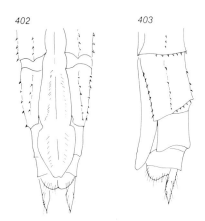

Figs. 402–403: *Crocothemis sanguinolenta arabica* (Schneider, 1982), female,
terminalia and vulvar scales, ventral and lateral views

Crocothemis erythraea (Brullé, 1832)

Figs. 389. 391–392. 395. 400–401

Libellula erythraea Brullé, 1832:102.
Libellula ferruginea Fabricuis, 1793:380. Schneider, 1845:111.
Crocothemis erythraea —. Brauer, 1868:737; Selys, 1887:22; Sowerby, 1917:10; Morton, 1924:42; Schmidt, 1938:148; Dumont, 1977b:158; Waterston, 1980:64; Schneider, 1981a:140; Lohmann, 1981:114.

Type Locality: Messenia, Greece.

Male

Labium ochraceous to reddish. Face, frons, vertex orange to red.
Thorax brown to red. Legs ochraceous or brown.
Wings hyaline, venation reddish. An amber patch at base of hind wing, almost reaching an_1, and down almost to anal angle. Pt long, yellow or brown.
Abdomen brown in tenerals, bright scarlet in mature specimens. A black carinal stripe may occur on S_9, sometimes partially extending to S_8 and S_{10}. Dorsal and lateral carinae set with black spines. 17–22 spines on S_5. Anal appendages red. Accessory genitalia: lam. ant. deeply hollowed-out, squared. Hamuli with inner branch not visible in lateral view, bifid apically. Genital lobe large, angular or rounded.
Female

Coloured as teneral male, yellow when immature, brown or olivaceous when mature.
Vulvar valvulae erect, triangular, pointed posteriorly, but rather short, not extending beyond S_9. A weakly developed basal, lateral swelling points posteriad.
Measurements (mm): Male. Total length 37–43; abdomen 21–28. *Female.* Total length 36–42; abdomen 21–28.
Distribution: Africa, Southern Europe, Asia Minor, and the Levant. In South-East Asia occurs *C. servilia* (Drury), which is closely related to *C. erythraea*. A dwarf subspecies from Iraq was described by Morton as *C. erythraea chaldaeorum. C. erythraea* is very common in Israel, on various types of waters. March–October.
Israel & Sinai (Locality records): Lake Hula (1), Hulata (1), Sedé Nehemya (1), Gonén (1), Ma'agan Mikha'él (4), 'Akko, (4), Nahalal (5), Deganya (7), Tabigha (7), Bet She'an (7), Tel Aviv (8), Rosh Ha'Ayin (8), Hadera (8), Qiryat 'Anavim (11), 'En Gedi (13), 'En Avedat (17), Et Tur (23), Gebel Katharina (22).

Crocothemis servilia (Drury, 1770)

Figs. 390. 393. 398–399

Libellula servilia Drury, 1770:112.
Crocothemis servilia —. Brauer, 1868:737; Lohmann, 1981:113; Schneider, 1985b:82.

Type Locality: "China".

Male

Coloured as *C. erythraea* from which it differs only in the characters stated in the key.

Female

Coloured as *C. erythraea* from which it differs in the characters stated in the key.

Dimensions: Japanese specimens are large, up to 5 mm. Usually longer in total length than *C. erythraea* but examples from continental Asia seem not to differ from it in size.

Distribution: Tropical and subtropical Asia. The ranges of *C. erythraea* and *C. servilia* meet and overlap in Anatolia, possibly Iraq, and in Jordan (and Syria?).

Locality records: The only record from the Levant so far is given by Schneider (1985b); oasis of El Azraq, Jordan, 1 male, 27.V.1980.

Crocothemis sanguinolenta arabica Schneider, 1982

Figs. 394, 396–397, 402–403

Libellula sanguinolenta Burmeister, 1839:859.
Crocothemis sanguinolenta —. Brauer, 1868:737; Ris, 1911:535; Morton, 1924:42.
Crocothemis sanguinolenta arabica Schneider, 1982:25.

Type Locality: Cape of Good Hope, South Africa.

Male

Face, legs and body yellow to ochraceous, later turning bright red. Sides of synthorax more brownish.

Wings hyaline, in old specimens with a tendency to become smoky near the apices. Venation reddish. Pt reddish-brown. Hind wing with basal amber extending to an_1 or an_2 and down to anal angle.

Abdomen bright scarlet. Incomplete black lines occur on dorsal and lateral carina of abdomen. Lateral carina with about 10–12 spines per segment on S_{4-6}. Accessory genitalia: lamina and hamuli as in *erythraea*, genital lobe elongate.

Female

Coloured as the immature male, but becoming darker at old age, and black stripes on dorsal and lateral carinae of abdomen more prominent than in the male. Vulvar valvules very long, projecting beyond S_{10}.

Distribution: Afrotropical.

Locality records: Morton (1924) mentions 3 males in Ris' collection from the "Dead Sea", June 1918. I found specimens on Wadi Mujib (= Naḥal Arnon) in June 1978, while 3 males are in the collection of Tel Aviv University, labeled Arnon River (13), 17.VIII.1941 (2 ♂ ♂) and 5.VII.1941 (1 ♂) (leg. Bytinsky-Salz.)

The specimens seen were all rather small, with total length 28–31mm; abdomen 18–20 mm.

Genus BRACHYTHEMIS Brauer, 1868

Verh. zool.-bot. Ges. Wien, 18:367

Type Species: *Libellula contaminata* Fabricius, 1793.

Medium-sized dragonflies, with rounded frons, small prothoracic hind lobes, robust synthorax. Legs rather long. Hamuli of male with long hook. S_8 in female not foliated. Vulvar valvules long. Wings short, broad, rounded. arc proximal to an_2. 6–7 an; last an incomplete. R_3 very slightly curved. Discoidal field with parallel margins throughout its course. d in forewing free or crossed, in hind wing entire. Membranula rather large.

Distribution: Africa and Asia.

Two regional species.

Key to the Species of Brachythemis
(Figs. 25, 404–413)

1. Males: wings traversed by a broad black band with base at N and top 1–2 cells proximal to Pt. Females rarely with wings banded, but with Pt bicolorous (base clearer than tip). Male genitalia: hamuli with outer branch depressed, inner branch a strong, upright hook. Genital lobe set with a small number of strong stiff hairs. Female: valvules triangular, apically rounded, with V-shaped invagination. **Brachythemis leucosticta** (Burmeister)
    Males: brown wing spot extending from wing base to N or 1–2 cells beyond N. Females with Pt uniformly yellow or brown. Male genitalia: hamuli with outer branch swollen, rounded, and inner branch an oblique hook. Genital lobe set with numerous fine long hairs. Females: valvules triangular, apically pointed, with U-shaped invagination in the middle. **Brachythemis fuscopalliata** (Selys)

Figs. 404–405: *Brachythemis leucosticta* (Burmeister, 1839)
404. teneral male; 405. aged female

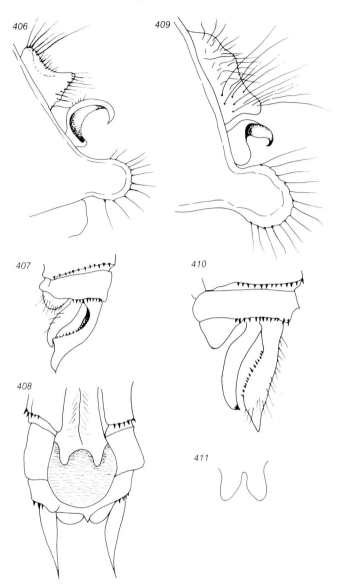

Figs. 406–411: *Brachythemis* spp.

406. *B. leucosticta* (Burmeister, 1839), male accessory genitalia, lateral view;

407. the same, male, anal appendages, lateral view;

408. the same, female, terminalia and vulvar area;

409. *B. fuscopalliata* (Selys, 1887), male, accessory genitalia, lateral view;

410. the same, male appendages, lateral view; 411. the same, female, vulvar lips

Figs. 412–413: *Brachythemis fuscopalliata* (Selys, 1887)
412. teneral male; 413. fully coloured male

Brachythemis leucosticta (Burmeister, 1839)

Figs. 404–408

Libellula leucosticta Burmeister, 1839:849. Selys & Hagen, 1850:310.
Trithemis unifasciata Rambur, 1842:108. Selys, 1887:23.
Cacergates leucosticta —. Kirby, 1889:306.
Brachythemis leucosticta —. Ris, 1909:27; Sowerby, 1917:10; Morton, 1924:42; Morton, 1929:60; Schmidt, 1938:148; Schmidt, 1954:83; St. Quentin, 1965:544; Dumont, 1977b:163; Waterston, 1980:64.

Type Locality: Durban, South Africa (but also cited are Egypt and Senegambia).

Male

Mouth parts, face, clypeus, frons pale yellow in teneral males, tending to blacken with age.

Synthorax yellow, with elaborate black markings (Fig. 404), progressively coated with black pruinescence. In very mature males, synthorax entirely velvet blue-black. Legs yellow marked with black. In old specimens, only back surface of tibiae yellow.

Wings: venation yellow. Pt clear yellow, apex darkened in teneral males, mainly because C and R are yellow in basal 2/3, black in top 1/3 of Pt. Membranula creamy in tenerals, dark grey in mature specimens. Wing spots barely visible in tenerals, soon expanding and darkening to form four complete wide brown patches, basally extending to N, apically to 2 cells proximal to Pt.

Abdomen yellow, with one mid-dorsal and two sub-lateral rows of black stripes. Basal black stripes on S_2, S_3, and on lateral carina of S_{4-5}. At later age, the abdomen turns darker, eventually becoming uniformly black. Appendages: superiors strongly curved, dark brown. Accessory genitalia: lam. ant. moderately raised, with short hairs. Hamuli prominent; outer branch depressed, slightly hollowed out, and the inner branch forming a strong, backwardly curved, upright hook. Genital lobe rounded, set with a small number of strong, stiff hairs.

Female

Strongly resembles the teneral male in head and synthorax. The hind wings often have basal amber, while Pt tends to darken with age, and become strongly bicolorous. In senescent females, it may become uniformly dark again. Wing spots either do not develop at all, or remain rudimentary. Abdominal markings differ from those in the male in that the mid-dorsal streaks are wider and the sublateral streaks narrower, split up into a series of elongated spots. Styli brownish, with black tips. Old females show a tendency towards melanism, though to a lesser degree than males. Vulvar aperture: valvules triangular, rounded apically, with a V-shaped invagination in the middle. Floor of S_9 broadly rounded posteriorly.

Measurements (mm): *Male.* Total length 36–40; abdomen 20–24. *Female.* Total length 35–39; abdomen 20–23

Distribution: Africa and the Levant; rare in Anatolia, but one of the most common dragonflies of the Jordan Valley.

Israel & Sinai (Locality records): Ḥula (1), Dan (1), Sedé Neḥemya (1), Rosh Pinna (1), Migdal (7), Tabigha (7), Naḥal Samakh (7), Bet She'an (7), Kinneret (7), Deganya (7), Tiberias (7), Bitanya (7), Bet Yeraḥ (7), Tel 'Amal [Nir Dawid] (7), Ashdot Ya'aqov (7), Ḥadera (8), Rosh Ha'Ayin (8), Et Tur (23), Sharm esh Sheikh (23).

Brachythemis fuscopalliata (Selys, 1887)

Figs. 25, 409–413

Trithemis fuscopalliata Selys, 1887:23.
Brachythemis fuscopalliata —. Ris, 1909–1919:335, 585; Fraser, 1917:282; Dumont, 1972:241; Dumont, 1977b:163; Schneider, 1981a:139.

Type Locality: Samava, lower Iraq.

Male

Mouth parts yellow, face green. With age, the clypeus and frons turn dark brown. Teneral males with synthorax dark olivaceous. In front, antehumerals visible as an inverted "V". Lateral sutures of synthorax clearly marked in black. Legs olivaceous, later brown, with exterior surface black.

Wings: Pt bright yellow. Membranula dark grey. A massive brown wing spot extends on all four wings from base to nodus or 1–2 cells beyond N in forewing, 3 cells beyond N in hind wing.

Abdomen brown, with end-rings broadly black, a black longitudinal carinal stripe, and a double sublateral stripe on each side. These markings quickly obliterated by a dark brown, uniform colour. Superior appendages ligth brown, strongly curved; inferiors almost black. Accessory genitalia: lam. ant. with numerous very long hairs; hamulus with external ramus swollen, rounded, and inner ramus an oblique, posteriorly directed hook. Genital lobe more deeply hollowed-out at base anteriorly than posteriorly, set with numerous long, fine hairs.

Female

Paler and smaller than the male, but otherwise with similar markings. Black striae on synthorax better visible and carinal stripe on abdomen wider. Styli yellow with black tips. Vulvar aperture: valvules broadly rounded, with a V-shaped invagination in the middle. Floor of S_9 rounded but strongly tapering towards its tip.

Measurements (mm): Male. Total length 36–39; abdomen 22–25. *Female.* Total length 31–34; abdomen 19–22.

Distribution: Iraq, south-east Anatolia, Syria, and northern Israel. Levantine capture data range from May to August. The species occurs in stagnant and slow-running waters.

Israel (Locality records): Lake Ḥula (1).

Genus ACISOMA Rambur, 1842

Insectes Neuropteres, p. 26

Type Species: *Acisoma panorpoides* Rambur, 1842.

Dragonflies of small size, coloured blue marked with black, and with a characteristic shape of the abdomen. Head small, eyes only just meeting. Pronotum with large hind margin, fringed with long hairs. Thorax narrow, small. Legs rather long, slim. Genitalia of male small and inconspicuous. Vulvar scales of female long, oval, projecting obliquely. Wings short, rather broad, reticulation rather open. d in forewing entire, in hind wing with base at arc, entire. Sectors of arc petiolated over a long distance. arc between an_1 and an_2. 7–9 an, the distal one complete or incomplete. Discoidal field beginning with 2 rows of cells, widening towards wing margin. Rspl with 1 row of cells. Membranula small. Pt relatively large.

Distribution: Asia and Africa.

One species.

Acisoma panorpoides ascalaphoides (Rambur, 1842)

Fig. 414

Acisoma ascalaphoides Rambur, 1842:29.
Acisoma panorpoides ascalaphoides —. Ris, 1909–1919:458; Andres, 1928:32.

Type Locality: "Madagascar".

Male

Labium whitish-ochre, sometimes with black medial lobe; labrum with fine black basal line and virgule. Clypeus and frons blue-green or pale blue. Frons with broad basal black band, continued laterally along the eyes.

Synthorax light blue, green-blue or pale green, with a complicated pattern of anastomosing black spots and streaks on sutures and on plates between sutures. Legs black with pale external stripes. Femora of first pair pale on flexor surface.

Wings hyaline, a trace of amber at base of hind wing. Pt pale yellow.

Abdomen markedly swollen on S_{1-4}, abruptly constricted on S_5, narrow and cylindrical on S_{6-10}. Superior appendage white on dorsum, black below. Appendix inferior black. A series of black spots on the dorsum of S_{1-5}; S_{6-7} black with lateral blue spot; S_{8-10} black. Accessory genitalia: lamina small; hamuli with exterior ramus long, narrow, apically rounded, internal ramus long, upright, hooked. Genital lobe very small, elongate.

Female

Similar to male. Face more greenish. Thorax green or olivaceous, blackish markings diffuse or even indistinct. Abdomen olivaceous or brown. End-rings, dorsal carina, and sides black. Vulvar aperture: valvules triangular, floor of S_9 broadly rounded.

228

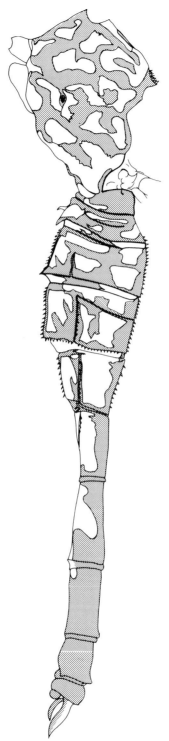

Fig. 414: *Acisoma panorpoides ascalaphoides* (Rambur, 1842), male,
markings on synthorax and abdomen

Wings: Pt dark yellow, between thick black veins.

Measurements (mm): *Male.* Total length 24–30; abdomen 16–22. *Female.* Total length 25–29; abdomen 16–21.

Distribution: Africa south of Sahara. Known in Egypt from the oases of Kharga and Baharieh; this species can therefore locally occur in Sinai.

Genus DIPLACODES Kirby, 1899
Trans. zool. Soc. Lond., 12:263

Type Species: *Libellula lefebvrei* Rambur, 1842 (or, in fact, its synonym *Libellula tetra*, over which *lefebvrei* has page priority).

Small dragonflies, with small head, short eye contact, rounded frons, a medium to large hind ridge of the pronotum, and rather small synthorax. Legs moderately long. Accessory genitalia of male not prominent. Vulvar scales of female elongate. Wing short, wide, with rather open venation. d in forewing only slightly distal to d in hind wing. Sectors of arc shortly fused in forewing, this fusion longer in hind wing. arc proximal to an_2. 6–9 an, the last one incomplete. d in forewing composed of 2–3 cells, in hind wing entire. R_3 nearly straight. Rspl with 1 row of cells. Discoidal field with 2 rows of cells, only slightly expanding towards the wing margin.

Distribution: Asia, Africa, Southern Europe.

One regional species.

Diplacodes lefebvrei (Rambur, 1842)
Figs. 415–416

Libellula lefebvrei Rambur, 1842:112.
Libellula morio Schneider, 1845:112.
Libellula flavistyla Rambur, 1842:117.
Diplacina flavistyla —. Selys, 1887:22.
Diplacodes lefebvrei —. Kirby, 1890:42; Morton, 1924:42; Schmidt, 1938:148; Schmidt, 1954a:68; St. Quentin, 1965:544; Dumont, 1977b:163; Waterston, 1980:63; Schneider, 1981a:140.

Type Locality: Oasis of Baharieh, Egypt.

Male

Mouth parts, face, frons, vertex black; frons above with violet sheen.

Synthorax and legs black at maturity, yellow in tenerals.

Wings hyaline, hind wing with basal yellow (teneral) or deep brown (mature) spot. Membranula white or greyish. Pt yellow to dark brown.

Abdomen slightly swollen at base, black with pale yellow lateral spots: a couple on S_{2-4}, becoming more like striae on S_{5-7}. An elongate spot at base of S_8, small spots on

S₉. All these, however, obliterated by black pruinosity at maturity. Anal appendages yellow. Accessory genitalia: lam. ant. low, with U-shaped excavation; hamulus a rectangular block with broad, rounded apex and with smaller inner upturned hook. Genital lobe elongate, tip broader than base, rounded.

Female
Like the teneral male, with mouth parts yellow, frons above with a black basal line. Synthorax and abdomen yellow, marked with black. Styli yellow. Vulvar aperture simple, only slightly produced, very shallowly excavated medially. Floor of S₉ broadly rounded, slightly produced over S₁₀.

A small species, but fairly variable in size.

Measurements (mm): *Male.* Total length 25–34; abdomen 16–25. *Female.* Total length 20–33; abdomen 15–24.

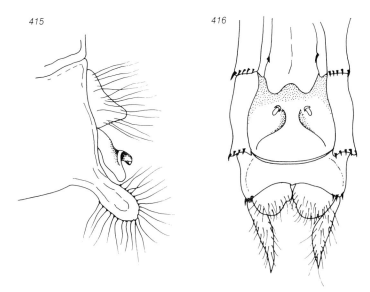

Figs. 415–416: *Diplacodes lefebvrei* (Rambur, 1842)
415. male, accessory genitalia, lateral view;
416. female, terminalia and vulvar lips (scales)

Distribution: The whole of Africa, reaching the Iberian Peninsula in the west and Anatolia in the east. Common in the Levant.
Israel & Sinai (Locality records): Lake Ḥula (1), Ḥadera (8), ʿEn Gedi (13), Wadi Tayiba (23).

Type Species: *Libellula aurora* Burmeister, 1839.

Small to moderate-sized dragonflies, with abdomen red, or black with yellow markings, often pruinose. Head small, eye contact short. Pronotum with small, rounded hind lobe. Synthorax small to moderately large. Legs rather long and slim. Abdomen variable, slender or depressed in males, cylindrical in most females. Male accessory genitalia prominent. Hamuli always with a strong apical hook. Females without foliate expansions on S_8; vulvar valvules very small. Wings moderately long to very long, fairly wide. d in forewing narrow or wide, situated slightly distal to d in hind wing. arc proximal to an_2. Sectors of arc fused at their origin. Last an usually incomplete. 8–15 an. d in forewing traversed, in hind wing usually entire. Discoidal field composed of three rows of cells throughout. R_3 generally nearly straight. Rspl with two rows of cells. Pt small or large. Membranula medium-sized.

Distribution: Oriental and Afrotropical regions, with two species reaching Southern Europe.

Three species are regional, while a fourth one should be expected in Sinai. In his revision of this genus, Pinhey (1970) has shown that the internal genital apparatus of the female (the bursa copulatrix) is rich in diagnostic characters. Unfortunately, this study has remained an isolated example so far.

Key to the Species of Trithemis
(Figs. 26, 417–430)

1.	Male	5
–	Female	2
2.	Vulvar valvules medially emarginate, deeply U-shaped	3
–	Vulvar valvules very shallowly emarginate, almost straight	4
3.	Abdomen brownish-olivaceous, with narrow sublateral black streaks, not broadly confluent towards top of segments. Membranula white to pale grey. Base of all wings usually with broad amber, extending to an_4 and beyond tip of d in hind wing.	

Trithemis kirbyi Selys

– Tenerals: abdomen yellow, marked with broad black streaks, expanding towards top of segments and confluent across dorsum; dorsum of S_{9-10} largely or entirely black. In fully mature specimens, most of thorax and abdomen blue-black pruinescent. Membranula dark brown. Base of hind wing with small, deep ochre-coloured spot.

Trithemis festiva (Rambur)

4. Abdomen at least slightly depressed, with segments 4–5 about twice as long as wide. Sides of abdomen not or only weakly and diffusely marked with black. Carinal black markings on S_{1-2} and S_{8-9} only. **Trithemis annulata** (Palisot de Beauvois)

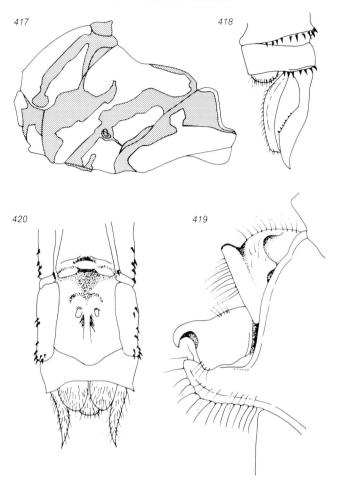

Figs. 417–420: *Trithemis annulata* (Palisot de Beauvois, 1805)
417. synthorax; 418. male terminalia, lateral view;
419. male accessory genitalia, lateral view;
420. female terminalia and vulvar lips, ventral view

–   Abdomen cylindrical, with S_{4-5} about three times as long as wide. Sides of abdomen broadly marked with black on S_{3-10}, these streaks widening towards apex of each segment, often broadly confluent across their dorsum, and with or without carinal stripe across dorsum of abdomen. S_{8-9} usually largely black.　　　**Trithemis arteriosa** (Burmeister)

5.　Hamuli triangular, gently tapering towards apex, apically hooked. A broad amber wing spot, roughly between wing base and midway to N on all four wings.

Trithemis kirbyi Selys

–   Hamulus abruptly constricted at about 2/3 of its length, with sickle-shaped apex. A wing spot at the base of the forewings　　　　　　　　　　　6

6.　Wing spot on hind wing rather small and dark, brownish. At maturity, most of thorax and abdomen blue-black. Tenerals yellow and black, marked as the female. Frons blue metallic.　　　　　**Trithemis festiva** (Rambur)

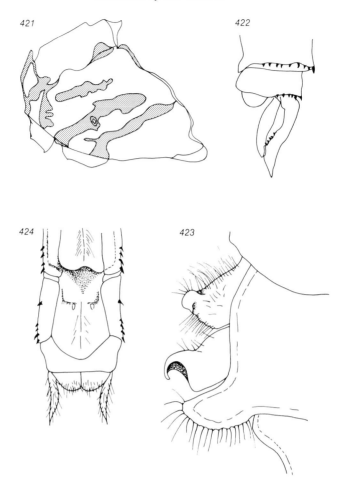

Figs. 421–424: *Trithemis arteriosa* (Burmeister, 1839)
421. synthorax; 422. male terminalia, lateral view;
423. male accessory genitalia, lateral view;
424. female terminalia and vulvar lips, ventral view

–   Wing spot on hind wing clear, amber-coloured. Most of the body bright red, eventually covered by a purple sheen, or marked with black at the sides only. Frons yellow or with purplish metallic sheen   7

7.   Abdomen depressed (S_{4-5} only twice as long as wide), yellow with diffuse carinal black markings in tenerals, bright red coated with purple pruinosity in adults. Frons yellow with narrow basal black bands in tenerals, with purple sheen in adults.

Trithemis annulata (Palisot de Beauvois)

–   Abdomen slender and cylindrical (S_{4-5} about three times as long as wide), yellow, in tenerals marked with black as in the female, bright red in adults, never with purple sheen. Frons yellow, with rather broad black basal band.   **Trithemis arteriosa** (Burmeister)

234

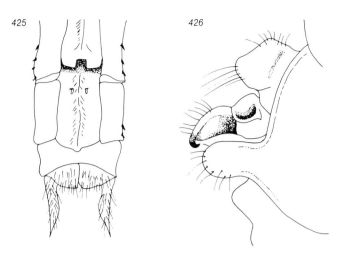

Figs. 425–426: *Trithemis kirbyi* Selys, 1891
425. female terminalia and vulvar lips, ventral view;
426. male accessory genitalia, lateral view

Fig. 427: *Trithemis kirbyi* Selys, 1891; male

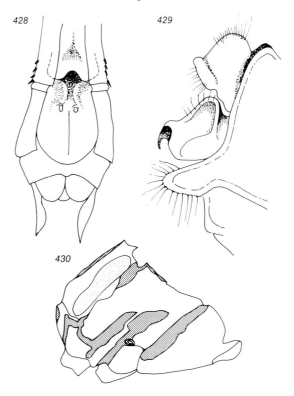

Figs. 428–430: *Trithemis festiva* (Rambur, 1942)
428. female terminalia; 429. vulvar lips, lateral view;
430. synthorax

Trithemis annulata (Palisot de Beauvois, 1805)

Figs. 417–420

Libellula annulata Palisot de Beauvois, 1805:69.
Libellula rubrinervis Selys, 1841:244.
Libellula haematina Rambur, 1842:84 (pars).
Trithemis ramburii Kirby, 1890:19.
Trithemis scortecci Nielsen, 1935:375.
Trithemis annulata scortecci —. El Rayah & El Din Abu Shama, 1978:81.
Trithemis selika Selys, 1867:2. Navas, 1911:3.
Trithemis annulata —. Morton, 1924:43; Schmidt, 1938:148; Valle, 1952:6; St. Quentin, 1965:544; Dumont, 1977b:164; Waterston, 1980:64; Schneider, 1981a:141.

Type Locality: Oware, Nigeria.

Male

Mouth parts and face ochraceous. Frons with shallow median groove, a metallic purple sheen on top in fully mature individuals. In some specimens, the labrum turns black with age.

Synthorax brownish, but pruinose dorsally, bright purple. Sides with black bands on the suture. Legs black, femora of first and second pairs yellow on flexor surface.

Wing venation red. Pt from dark yellow to rusty in colour. Forewing without basal amber. Hind wing with an amber spot extending from an_1 to the basal cells of the anal field. Membranula white or grey. Forewing with 10–11 an.

Abdomen somewhat depressed, red, coated with purple pruinescence at maturity. S_2 with traces of dorsolateral black stripes; S_{3-7} uniformly red; S_{8-9} with mid-dorsal black stripes on carina; S_{10} red. Anal appendages brown. Accessory genitalia: lam. ant. slightly produced, not bifid, bearing some rather short hairs. Hamuli robust, with broad base, constricted before their apex to produce a sickle-shaped, backturned hook. Genital lobe somewhat elongate, with apical and posterior hairs. Vesica spermalis compact, with rather short glans.

Female

Mouth parts and face entirely yellow; base of labrum with two isolated black spots. Frons with narrow black basal band. Synthorax with lateral black marking more expanded than in the teneral male, streaks often contiguous. Basal amber on hind wing generally more reduced than in male. Abdomen marked with black as in teneral males. Styli brown. Vulvar lips with shallow central depression. Bursa with dorsal sterigma bifid anteriorly, forming two twisted, divergent, tapering branches. Ventral sterigma ribbed, somewhat fan-shaped.

Measurements (mm): *Male.* Total length 32–38; abdomen 20–23. *Female.* Total length 32–35; abdomen 20–22.

Distribution: Africa, the main Mediterranean islands, S. Spain, S. Italy, Greece, S. Anatolia, the Levant. Fairly common and widespread in Israel and Sinai.

Israel & Sinai (Locality records): Lake Hula (1), Montfort (1), Qiryat Shemona (1), Malaha (1), Kefar Hittim (2), Haifa (3), 'Akko (4), Ma'agan Mikha'él (4), Nahalal (5), Tiberias (7), Deganya (7), "Beth Gordon", Deganya A (7), Umm Junni (7), Bitanya (7), Ashdot Ya'aqov (7), Bet She'an (7), Hadera (8), Rosh Ha'Ayin (8), Nabi Rubin (9), Bethlehem (11), Wadi Qilt (13), Wadi Fari'a (13), Massada (13), Abu Rudeis (23). Also recorded from Wadi Mujib (13), Jordan.

Trithemis arteriosa (Burmeister, 1839)

Figs. 421–424

Libellula arteriosa Burmeister, 1839:830.
Libellula conjuncta Selys, 1840:121.
Trithemis arteriosa syriaca Selys, 1887:23.

Trithemis arteriosa —. Morton, 1924:43; Andres, 1929:9; Schmidt, 1938:148; St. Quentin, 1965:544; Dumont, 1977b:164; Waterston, 1980:65.

Type Locality: Port Natal, South Africa.

Male

Labium yellow or reddish, labrum entirely brown-red or red, with finest black fringe. Clypeus reddish-brown. Frons and vertex reddish, with slight purplish reflection. A broad black line between frons and vertex. Occiput brown.

Synthorax brown or reddish, with dorsal purple sheen. Four lateral black bands of variable width (Fig. 421). Legs black, inner surface of femora 1 and 2 brown.

Wings: venation bright red. Pt red, with broad black posterior and anterior borders. Basal amber present in forewings and hind wings, but variable in extent, sometimes reduced to a trace, sometimes reaching d, especially in hind wing. Membranula grey.

Abdomen thin, bright red with lateral black markings variable with age and with population. Superior appendages brown, black at maturity, but usually some red left near the base. Inferiors red, with black margins and tips. Accessory genitalia: hamulus with robust base, abruptly constricted before its apex, with strong sickle-shaped, posteriorly directed hook. Genital lobe long and narrow, slightly bent over apically, hairy. Anterior lamina triangular in side view, moderately developed. Vesica spermalis short and broad, with large apical segment but no flagellum.

Female

Mouth parts and face, including frons, entirely yellow. Anteclypeus greenish. Vertex yellow, separated from frons by massive black stripe. Occiput yellow. Synthorax greenish-yellow or light brown, with faint antehumerals and lateral stripes as in male. Legs black; femora of first and second pair yellow on flexor surface. Wing venation brown; Pt brown, anterior and posterior borders black. Basal wing amber reduced to small spots. Styli black. Vulvar valvules with shallow invagination, but usually deeper than in *T. annulata*. Bursa copulatrix with dorsal sterigma narrowed anteriorly, well sclerotized, anteriorly with two detached, curved or angled, divergent, hirsute branches. Ventral plates curved.

Measurements (mm): *Male*. Total length 35–38; abdomen 22–26. *Female*. Total length 32–36; abdomen 20–24.

Distribution: The whole of Africa, where it is one of the most common Anisoptera. Also in Saudi Arabia, Iran, Iraq, the Levant. February–October.

Israel & Sinai (Locality records): Montfort (1), Haifa (3), Mt Carmel (3), Gesher (7), Ashqelon (9), Bethlehem (11), Aqua Bella (11), 'En Gedi (13), Timna' (14), 'En Meroha (17), Wadi 'Auja (13), 'En Mor (17), 'En Avedat (17), Wadi Nasb (22), Gebel Katharina (22), Wadi Hibran (22), Wadi Feiran (22), Wadi Isla (22), Wadi Talh (22), Wadi Tala (22), Sharm esh Sheikh (23), 'Ein el Furtaga (23).

Trithemis kirbyi Selys, 1891

Figs. 425–427

Trithemis aurora Kirby, 1886:327 (pars).
Trithemis kirbyi Selys, 1891:465. Pinhey, 1970:76; Waterston, 1980:66.

Type Locality: "India".

Male

Labium, labrum, clypeus, frons, vertex yellow with reddish sheen. Occiput deep yellow.

Synthorax olivaceous, with brown or reddish sheen. No dark carinal or antehumeral bands. Humeral suture with black streak on upper third. Black on sides of synthorax limited to Su_2 and lower sector of meso-metathoracic suture. Legs reddish-brown, extensor surface of femora black.

Wings: venation red, Pt short, dark brown, almost black on upper surface of wing, clear yellow fringed by thick black veins on inferior wing surface. A large amber spot at the base of all wings, modally extending to an_{4-5}, about midway between wing base and nodus. In hind wing, the spot may not extend to the wing base in the anal area. Often, distinctly clearer cells occur in the anal and basal wing zone.

Abdomen uniformly red, appendages red or deep brown. A short black carinal stripe on S_9. Accessory genitalia: lam. ant. swollen, notched at apex; hamulus a long triangle gently tapering towards its hooked apex. Genital lobe long and rounded. Vesica spermalis robust, with small apical segment and slender flagella.

Female

Mouth parts, clypeus, frons, vertex yellow or greenish-yellow. Wings as in the male, but wing spots variable. In forewing, they are often reduced to narrow streaks in the subcostal and cubital space. In India, a truly homochrome form exists alongside a form with the wing spots completely reduced, while in Africa homochrome forms are rare, except in Nigeria where they are dominant. Abdomen more robust than in the male, triquetral or somewhat depressed, olivaceous or brownish. S_1 dorsally black, sublateral black stripes on S_{4-9}, and mid-dorsal carinal black marks on S_{8-9}. S_{10} largely black. Styli brown or black. Vulvar valvules deeply hollowed-out medially, with lateral lips swollen. Bursa copulatrix: central sterigma folded and bursal arms slender.

Measurements (mm): *Male.* Total length 30–33; abdomen 19–22. *Female.* Total length 31–34; abdomen 19–23.

Distribution: Typical *T. kirbyi* (with wing fascia extending to an_{3-4}) occurs in India; in Africa a form is found with larger wing spots (extending to an_{6-7}), which is considered to be a separate subspecies (*T. kirbyi ardens* Ferstäcker). However, populations in the Maghreb countries, in the Sahara, and in Saudi Arabia present an intermediate wing spot (as described above). Until the status of all these forms is clearly settled, no subspecific status can be accorded to the North African and Middle Eastern form. *T. kirbyi* has not been recorded from Sinai, but Waterston (1980)

239

reports it from the Red Sea mountains near Jeddah, so that its occurrence further north appears reasonable (the species lives on the southern slopes of the Atlas Mountains, hence no climatic barriers exist).

Trithemis festiva (Rambur, 1942)

Figs. 428–430

Libellula festiva Rambur, 1842:92.

Trithemis festiva —. Brauer, 1868:736; McLachlan, 1899:301; Schmidt, 1954a:83; Valle, 1952:6; St. Quentin, 1965:544; Dumont, 1977b:163.

Trithemis festiva rhodia —. St. Quentin, 1964b:660 *nomen nudum.*

Type Locality: Bombay, India.

Male

Labium dark brown, labrum dark olivaceous to black, with base brown. Anteclypeus black. Postclypeus and frons dark olivaceous brown. Frons deep metallic violet above. Vertex metallic violet. Occiput dark brown.

Synthorax black, coated with purplish pruinescence. Legs black.

Wings hyaline with a cognac-coloured wing spot at the base of the hind wing. Membranula dark, with paler margins. Pt black.

Abdomen slim and triquetral in cross-section, black with bluish pruinescence on S_{1-3}. Anal appendages black. Teneral males have the synthorax and abdomen yellow, very copiously marked with black. Accessory genitalia: lam ant. rather small; hamuli massive, tip constricted, with small sickle-shaped hook. Genital lobe elongate, tapering to apex.

Female

Labium pale brown, middle lobe black. Labrum, face and frons dirty yellow, changing to dark brown on top of frons (non-metallic). Synthorax olivaceous or yellow, with a broad black humeral stripe, a carinal stripe that tapers to a point at the interalar sinus, an inverted Y-shaped stripe on mesepimerum, a narrow stripe on Su_2, and a short oblique stripe extending from Su_2 backwards across metepimerum. Legs black, anterior femora yellow on flexor surface. Wings as in male, but in adults the wing tips are broadly darkened, to as far as the proximal edge of Pt. Abdomen cylindrical, yellow, longitudinally marked with black on mid-dorsum on the end-rings of the segments and sublaterally. All these stripes are confluent at the ends of all segments, isolating yellow spots between them and eventually obliterating all yellow colour on S_{7-10}. Styli black. Vulvar scales narrowly but deeply emarginate. Floor of S_9 rounded, elongate, produced over S_{10}.

Measurements (mm): *Male.* Total length 32–37; abdomen 22–27. *Female.* Total length 31–36; abdomen 21–25.

Distribution: India, Pakistan, Afghanistan, Iran, Anatolia, Cyprus, Samos, Rhodes. In the Levant it is found as far south as the Jordan Valley. May–November.

Israel (Locality records): Deganya (7), Rosh Ha'Ayin (8), Jerusalem (11).

240

Genus ZYGONYX Hagen, 1867

Verh. zool.-bot. Ges. Wien, 17:62

Type Species: *Zygonyx ida* Selys, 1869.

Moderate- to large-sized libellulids, with strong metallic sheens. Head large, eyes well fused, frons rounded. Pronotum with small, depressed hind lobe. Synthorax robust. Legs long. Abdomen robust and constricted at S_3 in males, cylindrical thereafter. S_8 of females without foliations. Vulvar valvules small. Wings long. N closer to apex than to base. arc between an_1 and an_2. Last an incomplete. R_3 bisinuous. Discoidal field with 3–4 rows of cells, parallel-sided Rspl with 1–2 rows of cells. d in forewing crossed. Pt rather short.

Distribution: A riverine genus found in both the Afrotropical and Oriental regions. One regional species.

Zygonyx torrida torrida (Kirby, 1889)

Fig. 431

Pseudomacromia torrida Kirby, 1889:299. Morton, 1924:43.
Zygonyx torrida —. Schmidt, 1938:148.

Type Locality: Sierra Leone, western Africa.

Male

Labium entirely black or with yellow sides. Labrum, postclypeus dark; anteclypeus olivaceous. Most of frons and vertex metallic black, with violet sheen.

Pronotum black. Synthorax black with metallic sheen and yellow markings variously covered with grey pruinosity in mature specimens. Legs black.

Wings hyaline, no basal wing spot. Venation dark brown or black. Membranula whitish. Pt black. Usually 2 rows of cells in Rspl. Last an in forewing usually complete. Abdomen black with yellow spots; those on S_{2-3} paler than on rest of abdomen. Anal appendages black. Accessory genitalia: lam. ant. relatively small, conical; hamulus with robust base and sickle-shaped apex. Genital lobe rounded.

Female

Very much like the male, but abdomen more massively built. Wings often suffused with saffron. Frons with less metallic black in front than in male. Vertex reddish or ochraceous.

Measurements (mm): Male. Total length 50–56; abdomen 35–38. *Female.* Total length 52–59; abdomen 38–41.

Distribution: Most of Africa, with relicts in the Maghreb countries. A distinct subspecies (*Z. torrida isis* Fraser) occurs in India. A stream-dweller that favours rapids and waterfalls. Flight period from March to October.

Israel (Locality records): Bet Qeshet (2), Deganya (7), 'Ein Duyuk (13), 'En Gedi (13), 'Ein es Sultan (13), Jericho (13), Wadi 'Auja (13), Wadi Fari'a (13), Wadi Qilt (13).

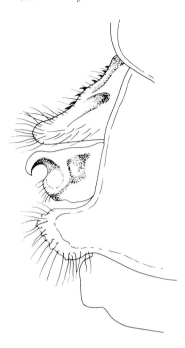

Fig. 431: *Zygonyx torrida torrida* (Kirby, 1889), male,
accessory genitalia, lateral view

Genus RHYOTHEMIS Hagen, 1867
Stettin. ent. Ztg, 28:232

Type Species: *Libellula phyllis* Sulzer, 1776.
Medium-sized libellulids with small head, short eye contact, small rounded frons.
Pronotum with hind lobe small and depressed. Synthorax small. Legs long, slender,
with numerous fine tibial spines. Abdomen short and slender. Females without
foliations on S_8. Valvules small. Wings, especially hind wings, broad with coloured
metallic spots. d in forewing free, large and rather broad, in hind wing 3–4 cells distal
to that in forewing; sectors of arc fused at their origin. arc proximal to an_2, usually
closer to an_1. Last an incomplete. R_3 and Cu_2 straight or only weakly curved. Discoidal
field composed of 3 rows of cells, more or less parallel throughout. Pt small.
Membranula large.
Distribution: Afrotropical, Oriental and Australian regions, extending to the Pacific
islands.
One species is regional.

242

Rhyothemis semihyalina syriaca (Selys, 1849)

Figs. 432–435

Libellula syriaca Selys, 1849:115 (race of *Libellula separata* Selys).
Libellula syriaca —. Selys & Hagen, 1850:305.
Rhyothemis hemihyalina Selys, 1887:8.
Rhyothemis syriaca —. Kirby, 1890:6.
Rhyothemis semihyalina (Desjardins, 1835:3). Morton, 1924:44; Schmidt, 1938:149.
Rhyothemis semihyalina syriaca —. Dumont, 1975:3.

Type Locality: "Syria", but almost certainly Lake H̲ula area.

Male

Labium and face yellow or reddish-brown. Labrum black. Frons and vertex metallic violet. Body black. Sides of thorax and abdominal segments 1–4 dark rusty or black with metallic sheen, in older specimens covered with pale pruinosity.

Wings hyaline, hind wing with broadened base, covered by an extensive metallic black spot, almost reaching to last an, leaving only a narrow hyaline fringe along the posterior wing margin. Pt very small, usually black.

Appendages black. Accessory genitalia: lam. ant. massive, trapezoidal in side view; hamuli apically hooked; genital lobe moderately developed.

Fig. 432: *Rhyothemis semihyalina syriaca* (Selys, 1849), male (paratype)

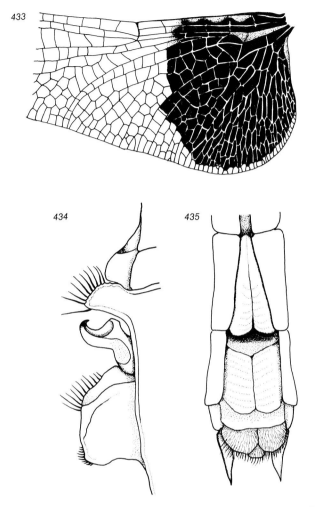

Figs. 433–435: *Rhyothemis semihyalina syriaca* (Selys, 1849)
433. hind wing;
434. male, accessory genitalia, lateral view
435. female, terminalia and vulvar lips, ventral view

Female
Very similar to male.

Measurements (mm): *Male.* Total length 29–32; abdomen 19–22. *Female.* Total length 28–30; abdomen 16–19.

Distribution: Restricted to Lake Ḥula area (1) where the last known specimen was captured in 1950. Thus, the subspecies is almost certainly extinct. The nominal subspecies is distributed over Africa south of the Sahara, and had a relict population in northern Algeria (Lake Oubeira). In Lake Ḥula area, specimens have been collected between May and August. The species has a dancing flight, much like a butterfly, and must therefore have been rather conspicuous.

Genus PANTALA Hagen, 1861

Synopsis Neuroptera N. Am., p. 141

Type Species: *Libellula flavescens* Fabricius, 1798.
Medium-sized libellulids, with a very large head, a small pronotum, and a robust synthorax. Legs long and slender. Wings long, hind wing very broad at base. d in forewing 2–3 cells distal to d in hind wing. Sectors of arc fused at origin; arc between an_1 and an_2; 13 an, the last one incomplete; d in forewing traversed. Discoidal field first 3 rows of cells, later 4–5, but contracted at wing margin. R_2 strongly bisinuous. Pt very short, longer in forewing than in hind wing. Abdomen cylindrical, dilated at base, constricted at S_3 in males. Anal appendages of males long and slender.
Distribution: Cosmopolitan, but most curiously very rare or lacking in Europe. One almost cosmopolitan species is regional.

Pantala flavescens (Fabricius, 1798)

Figs. 436–437

Libellula flavescens Fabricius, 1798:285.
Pantala flavescens —. Cabot, 1889:43; Selys, 1887:8; Morton, 1924:44; Andres, 1928:8; Dumont, 1977b:164.

Type Locality: "India".

Male
Labium ochraceous, with darkened median lobe. Labrum ochraceous or brown. Face, frons, and vertex yellow, sometimes suffused with red.
Pronotum ochraceous and black. Synthorax hairy, greenish-brown, rather transparent. Sides greenish, often with brown spots on the sutures. Legs dark brown or black.
Wings hyaline. Broad hind wing with basal amber spot; rarely wing tops also suffused with amber. Membranula white. Pt light reddish brown, longer in forewing than in hind wing.
Abdomen cylindrical, tapering posteriorly, yellow to deep brown, or even red, with black mid-dorsal carinal band. Anal appendages long, reddish, black at apices. Accessory genitalia: lam. ant. small, triangular in side views; hamuli long and slender, with apical hook slightly turned outwards. Genital lobe rounded.
Female
Habitus as in male, but body colours less bright. Vulvar valvules fairly deeply invaginated, U-shaped, with lips swollen.
Measurements (mm): *Male.* Total length 47–52; abdomen 30–34. *Female.* Total length 49–55; abdomen 32–37.

Fig. 436: *Pantala flavescens* (Fabricius, 1798), female

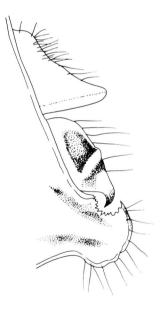

Fig. 437: *Pantala flavescens* (Fabricius, 1798), male,
accessory genitalia, lateral view

Distribution: Cosmopolitan, but extremely rare (if at all present) in Europe. A strong migrant, capable of covering continental distances, often encountered in deserts and on open sea.

Israel (Locality records): 'Ubeidiya (7; 13.X.1935) and 'En Gedi (13; 16.VIII.1957).

Genus UROTHEMIS Brauer, 1868
Verh. zool.-bot. Ges. Wien, 18:175

Type Species: *Urothemis bisignata* Brauer, 1868.

Moderately large species, with large head, wide frons and swollen vertex. Pronotum with hind lobe small; synthorax strongly built, with robust legs. Abdomen robust, depressed. Wings long, nodus situated about midway along their length. Hind wings with broadened base. d in forewing broad, situated somewhat distal to d in hind wing. Sectors of arc not or only shortly fused. arc proximal to an_2. 7–8 an, the last one complete. One row of cells in Rspl. Discoidal field with 3 rows of cells, not expanded at wing margin. Pt medium-sized. Membranula large. Male accessory genitalia: hamulus long and pointed. Female: S_8 not foliate, valvules well developed.

Distribution: Afrotropical and Oriental regions.

Urothemis edwardsi hulae Dumont, 1975
Figs. 438–443

Urothemis edwardsi Selys, 1849:124. Morton, 1924:44; Schmidt, 1938:149.
Urothemis edwardsi hulae Dumont, 1975:3.

Type Locality: Lake Hula, Israel.

Male

Labium yellowish; labrum yellow or reddish with blackened base. Anteclypeus olivaceous; postclypeus, frons, vertex dark green to black, partly with bluish sheen.

Pronotum, synthorax and abdomen blue-black in mature specimens, covered by bluish pruinosity which eventually reaches wing base. Legs black; femora of first pair brown on flexor surface.

Wings hyaline or slightly suffused with amber, especially near their tips. Pt yellowish. Base of hind wing with a small brown spot, surrounded by some amber, not normally extending into subcostal space on the one hand, of beyond the level of the basal cubito-anal cross-vein on the other hand.

A black band on the mid-dorsum of the abdomen remains visible throughout life. It expands sharply near the top of each segment. Anal appendages dark brown. Superiors, in side view, widened at about half their length, with inferior hook.

Fig. 438: *Urothemis edwardsi hulae* Dumont, 1975; female (paratype)

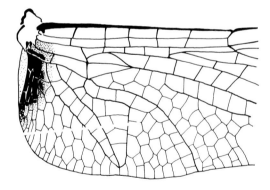

Fig. 439: *Urothemis edwardsi hulae* Dumont, 1975; hind wing

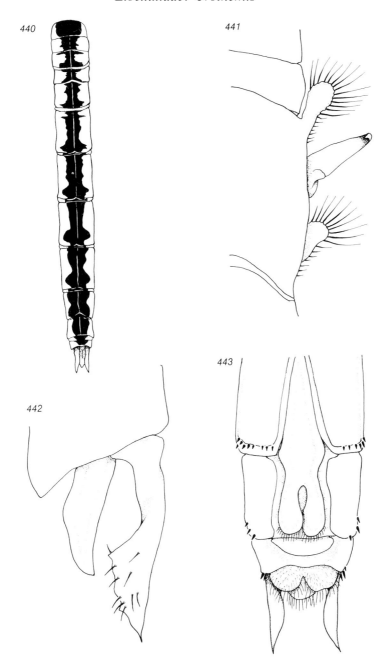

Figs. 440–443: *Urothemis edwardsi hulae* Dumont, 1975
440. male, abdomen; 441. male accessory genitalia, lateral view;
442. male terminalia, lateral view;
443. female terminalia and vulvar scales, ventral view

Teneral males differ in being paler, their abdomen being yellow or clear brownish, with a black longitudinal band. Accessory genitalia: lam. ant. very small, with distinctive long hairs. Hamuli elongate, triangular, tapering towards their pointed and outwardly curved apex.

Female

Paler than the male. Mouth parts and face light yellow or green. Thorax greyish, with yellow flanks, and black spots at the sutures. Wing spot on hind wings as in the male.

Measurements (mm): *Male.* Total length 42–47; abdomen 26–30. *Female.* Total length 42–45; abdomen 26–28.

Distribution: Restricted to Lake Ḥula (1) where specimens have been taken between June and November. A single female was collected at Jericho (13), 16.VIII.1942 (a migrant or a mislabeled specimen?). No specimens have been collected since 1952 (after the drainage of Lake Ḥula), and the subspecies is probably extinct, like *Rhyothemis semihyalina syriaca* with which it co-occurred. The nominal subspecies is widely distributed in Africa south of the Sahara, and was recorded once on Lake Oubeira, N.E. Algeria (loc. typ.).

Genus SELYSIOTHEMIS Ris, 1909
Catal. Collns Selys Longchamps, 9:37

The genus is monotypic.

Small dragonflies, with black non-metallic body-colours and colourless wings. Head relatively large; eyes broadly contiguous. Frons flattened in front, with deep median groove. Pronotum with small hind lobe. Synthorax small. Legs long and slender. Abdomen short, dilated at base, slightly constricted at S_3, then cylindrical to its tip. Wings short and broad, with open venation. d in forewing broad, entire; in hind wing situated at arc, entire. Sectors of arc separated from their origin in forewing, shortly fused in hind wing. arc between an_1 and an_2. 5–6 an, the last one complete. Discoidal field with 2 rows of cells at origin, expanded at wing margin. Rspl with one row of cells. Pt very small. Membranula medium-sized. Wing venation whitish. Male accessory genitalia: lam. ant. depressed, broadly arched. Hamuli broadly triangluar, apex with short curled hook. Genital lobe narrow, truncate. Female: sides of S_8 not foliate; vulvar valvules very small.

Selysiothemis nigra (Vander Linden, 1825)
Figs. 444–445

Libellula nigra Vander Linden, 1825:16. Selys, 1840:29; Selys & Hagen, 1850:65.
Urothemis nigra —. Selys, 1878:4; Selys, 1887:77.
Urothemis advena —. Selys, 1878:4; Selys, 1887:77.
Selysiothemis nigra —. Ris, 1897:48; Fraser, 1936:451; Schmidt, 1938:149; Dumont, 1977b:164;
  Schneider, 1981b:97.

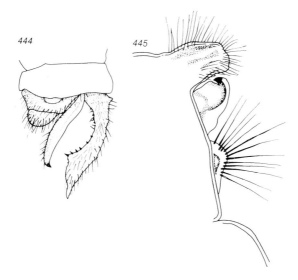

Figs. 444–445: *Selysiothemis nigra* (Vander Linden, 1825), male
444. terminalia, lateral view; 445. accessory genitalia, lateral view

Type Locality: Area of Naples, Italy.

Male
Head large, labium yellow, labrum dark brown or black. Face and vertex deep olivaceous. A black stripe present at the base of the frons. Synthorax hairy, dark olivaceous with black sutures, completely black in senescent specimens. Abdomen black, end-rings and sides dark brown. A greyish pruinosity develops on the abdomen and the base of the legs at maturity. Legs black. Appendages: superiors strongly arched, their apical part swollen, dark brown; inferiors black.
Teneral males differ in being paler: labrum yellow, synthorax greenish, abdomen pale green or brown with carinal black markings that widen to the top of each segment. Legs with basal half of the femora brown. Other characters as for genus.
Female
Coloured as the teneral male. Other characters as for genus.
Measurements (mm): *Male.* Total length 30–38; abdomen 21–26. *Female.* Total length 32–37; abdomen 22–25.
Distribution: An Irano-Turanian species that extends into North Africa, and descends deep into Sahara in its mountainous centre, much like *Orthetrum ransonneti*. The species is common in the eastern Mediterranean area, but becomes rare to exceptional in the west, where stable colonies are found on the Balearic Islands only (Comte-Sart, 1960). Records range between May and August.
Israel & Sinai (Locality records): Sedé Neḥemya (1), Rosh Ha'Ayin (8), Golan (18), Wadi Tayiba (22), Et Tur (23).

MORPHOLOGY OF THE LARVAL STAGES

Between the hatching from the egg and the emergence of the adult, dragonfly larvae undergo a number of moults, in some species interrupted by a larval diapause, and variable in duration between a few weeks and five, perhaps even six, years. The very first instar or prolarva is extremely short-lived, and has non-functional appendages. It moults within hours into the first active larval instar that immediately starts feeding. In a few cases only is the whole larval development known, and in many species nothing at all is known about the immature stages. Existing keys are therefore based on the terminal larval instar, which may differ considerably from the preceding instars. Identification of random samples of dragonfly larvae is therefore still a very cumbersome, if not impossible task.

In the Near East, in particular, a sizeable amount of work remains to be done, and the keys following this anatomical introduction must therefore be considered as a first approximation only. It is also quite likely that future approaches to dragonfly larval morphology will lead to the discovery of more reliable distinctive characters than those used today.

Head

Considerably less mobile than in the adult, but consisting of the same main parts. Compound eyes large but more depressed, ocelli present or absent. Occiput larger than in the adult, due to the existence of broad postocular extensions. Antennae relatively longer than in adults, composed of 3–7 segments, the modal number being 7. Clypeus and frons not yet individualized. Mouth parts raptorial as in the imago, except for the labium, which is differentiated into a characteristic prey-catching device, the mask. The mask is composed of a basal part, the submentum, connected by a hinge-joint to the apical prementum. The apex of the prementum is formed by a labial palpus and a movable hook. The mask is flat in Zygoptera, Gomphidae and Aeschnidae, concave in Libelluloidea. The prementum is set with a number of setae and spines. These structures are useful for identification down to species level. The shape of the mandibles, and the number and arrangement of teeth on the mandibles are also important. The left and right mandibles are asymmetrical.

Thorax

The prothorax is comparatively larger than in adults; the synthorax is fused. The legs may present a basal outgrowth, the supra-coxal armature that is used in the identification of aeschnid larvae. The wings are visible as alar furrows, extending backwards over the dorsum of the abdominal segments, and growing longer with every moult.

252

Abdomen

Ten distinct segments, and traces of an 11th one are present. S_1 is shorter than the subsequent ones that may or may not bear mid-dorsal spines. Spines may also occur at the posterior external angles of some or all segments in many Anisoptera. In Zygoptera, the abdominal apex bears three caudal lamellae – elongated, foliated, highly tracheolated structures with a respiratory function. Colour bandings, marginal setation or spinulation of these lamellae are diagnostic. In addition to these, larval Euphaeidae also have paired lateral abdominal gills. Anisoptera feature an anal pyramid composed of the following elements: a mid-dorsal epiproct (in males with a basal male appendix), a pair of lateral cerci (the future superior appendages), and a pair of ventro-lateral paraprocts.

The sex of a larva can be identified by the presence of a pair of small gonapophyses on the ventrum of S_2 in males, and by the conspicuous gonapophyses on the ventrum of S_{8-9} in the females of Zygoptera, Aeschnidae and Cordulegasteridae. As stated before, all anisopteran males also possess a male projection at the base of the dorsal surface of the epiproct.

Key to Suborders

1.	Larvae of elongated shape, slender, with abdomen terminated by three caudal lamellae.	**Zygoptera**
–	Larvae relatively robust, less elongated, with abdomen terminated by a compact anal pyramid composed of five appendages.	**Anisoptera**

Suborder Zygoptera: Key to Families

1.	Paired abdominal gills present on segments 2–8.	**Euphaeidae**
	Only one genus and species:	**Epallage fatime** (Fig. 446)
–	No abdominal gills	2
2.	Antenna: scapus half the length of the entire antenna. Prementum with a deep median cleft, about half depth of the entire prementum.	**Calopterygidae**
	Only one genus:	**Calopteryx** (Figs. 447–448)
–	Antenna: scapus less than 1/4 the length of the entire antenna. Prementum with median cleft narrow and much shorter than half the length of the prementum or absent	3
3.	Prementum with a narrow and short median cleft. Movable hook with setae.	**Lestidae**
–	Prementum without median cleft. Movable hook without setae	4
4.	Labium with 4 premental setae, arranged in 2 groups of 2 each, implanted on a horizontal line (Fig. 454). Pronotum with two medio-lateral bosses.	**Platycnemididae**
	One genus:	**Platycnemis** (Figs. 454, 455)
–	Labium with a variable number of premental setae, implanted along an arched line. Pronotum without lateral bosses.	**Coenagrionidae**

Family Lestidae: Key to Genera

1. Prementum abruptly constricted somewhat before half its length, parallel-sided thereafter (Figs. 451–453).   **Lestes**
                                                                                    (minus *L. viridis*)
–   Prementum triangular   2
2. Median cleft of prementum very shallow. Labial palpus deeply divided, with a strong medial hook and a wide outward portion, the later with inner hook and 10–12 crenations (Fig. 450).   **Lestes viridis parvidens**
–   Median cleft of prementum deeper. Outer part of divided labial palpus with strong inner hook and 2 smaller teeth along its free margin. (Fig. 449).   **Sympecma fusca**

Family Coenagrionidae: Key to Genera

This key is provisional, unsatisfactory in many respects and requires extensive future revision. Indeed, since numerous larvae in *Agriocnemis, Pseudagrion, Ceriagrion* and *Ischnura* still await discovery and description, modifications of generic diagnoses may be expected as these forms become known.

1. One pair of premental setae (Fig. 456).   **Pseudagrion***
–   More than one pair of premental setae   2
2. Maximum 3 pairs of premental setae (Fig. 457).   **Ceriagrion***
–   Minimum 3, usually at least 4 pairs of premental setae   3
3. Antenna with 6 segments (or 7th segment indistinct)   4
–   Antenna with 7 segments (Fig. 467)   5
4. Full-grown larva around 9 mm long. Movable hook on labial palpus much shorter than strongly developed internal end-hook.   **Agriocnemis**
–   Full-grown larva 25–30 mm long. Movable hook of labial palpus much longer than end-hook (Fig. 462).   **Erythromma**
5. Labium with 6–10 (3–5 pairs) premental setae (Figs. 459–461).   **Coenagrion**
–   Labium with 8–12 (4–6 pairs) of premental setae (Figs. 463–469).   **Ischnura**

* *Ceriagrion glabrum*, from Central Africa, has one pair of premental setae (Corbet, 1956), but does not occur in the Jordan Valley.

Suborder Anisoptera: Key to Families

1. Prementum flat; labial palpi not crenated except in *Caliaeschna microstigma*   2
   – Prementum concave; labial palpi crenated   3
2. Antenna composed of 7 segments. A couple of bosses at the base of the prothoracic legs (precoxal armature: Fig. 485).   **Aeschnidae**
   – Antenna composed of 4 segments (Fig. 474). No precoxal armature.   **Gomphidae**
3. Median lobe of prementum apically with a wide V-shaped, shallow cleft. Margin of labial palpi strongly serrated (Figs. 480–481).   **Cordulegasteridae**
   One genus and species: **Cordulegaster insignis**
   – Median lobe of prementum not cleft. Margin of labial palpi variably structured but never deeply serrated.   **Libellulidae**

Family Gomphidae: Key to Genera

1. Total length of full grown larva ca. 45 mm; larva robustly built, with broad abdomen (Figs. 470–471).   **Lindenia**
   – Total length of full-grown larva not exceeding 30–35 mm; larva fusiform, with abdomen more slender (Fig. 472)   2
2. Distal margin of prementum straight (Fig. 473).   **Gomphus**
   – Distal margin of prementum more or less curved   3
3. Prementum: distal margin slightly convex. Labial palpus with inner margin denticulate. Abdomen: lateral spines on S_{7-9}, and mid-dorsal spines on S_{2-9}.   **Onychogomphus**
   – Prementum: distal margin markedly convex. Labial palpus with inner margin smooth. Abdomen: lateral spines on S_{2-9}, small mid-dorsal spines on S_{2-3} (Figs. 475–479).   **Paragomphus**
3a. Apical segment of antenna very elongate. Distal margin of prementum medially pointed (Figs. 478–479).   **P. genei**
3b. Distal segment of antenna more stoutly built. Distal margin of prementum rounded (Figs. 476–477).   **P. sinaiticus**

Note
These key characters are based on non-regional species; since the larvae of the three regional *Onychogomphus* are still unknown, this may affect the value of the key-characters.

Family Aeschnidae: Key to Genera

1. Distal margin of prementum without median cleft, its central zone angularly protruding. Inner margin of labial palpus rather grossly crenated. Full-grown larva smaller than 35 mm (Figs. 482–486).   **Caliaeschna**
   – Distal margin of prementum with a median cleft, its central zone straight or convex. Inner margin of labial palpus smooth or finely crenated. Full-grown larva well over 35 mm long   2

2.   Eyes not notably flattened on dorsum of head, their median posterior margin protruding posteriorly (Fig. 496b). Posterior rim of head angular. Total length of full-grown larva less than 4.5–5 cm   3

–   Eyes flattened on dorsum of head, their posterior margin straight. Outline of head more or less rounded. Total length of full-grown larva 4.5–5 cm or more   4

3.   Cerci less than 2/3 the length of the paraprocts (Figs. 487–488, 495–496).   **Aeshna**

–   Cerci 2/3 of the length of the paraprocts (Fig. 489).   **Anaciaeschna**

4.   Maximum width of prementum less than 70% of its length (prementum long) (Fig. 492).

    Anax

4a.   Supra-anal lamina about as wide as long (Fig. 491).   **A. imperator**

4b.   Supra-anal lamina wider than long (Fig. 492).   **A. parthenope**

–   Width of prementum about 70% of its length (prementum short) (Fig. 494).

    Hemianax

Family Libellulidae: Key to Genera

Still undescribed are the larval stages of *Diplacodes lefebvrei* and *Acisoma panorpoides*, of several of the regional *Orthetrum*, of *Brachythemis fuscopalliata*, of several *Trithemis*-species, and of *Sympetrum decoloratum*. Further, *Urothemis edwardsi hulae* and *Rhyothemis semihyalina syriaca* became extinct before their larvae were discovered.

1.   Larvae with hairy aspect (Fig. 497). Eyes small, with well- developed postocular lobes, giving the head a robust, rectangular aspect. Legs relatively short, the anterior pair widened. Lateral spines on S_{8-9} weakly developed   2

–   Larval surface smooth, not woolly in aspect. Eyes large, with postocular lobes relatively small, so that the anterior and posterior borders of the head are more rounded. Legs long and slender, the anterior pair not flattened. Lateral spines on S_{8-9} often robust   3

2.   Mid-dorsal spines on abdomen small or absent. No such spines on S_8. Sides of head parallel (Fig. 499).   **Orthetrum**

–   Mid-dorsal spines on abdomen robust. Always a spine on S_8. Sides of head slightly convergent (Figs. 497, 498).   **Libellula**

3.   No mid-dorsal spines on abdomen   4

–   Mid-dorsal spines on at least some abdominal segments   5

4.   Total length around 20 mm. Lateral spines on S_{8-9} short, those on S_9 not more than 1/4 of the length of that segment (Fig. 500: prementum).   **Crocothemis***

–   Total length exceeding 25 mm. Lateral spines on S_{8-9} robust and long, those on S_9 longer than the segment itself (Figs. 503–504).   **Pantala flavescens**

5.   Labial palpus with 4–5 setae; premental setae 12–16 in two rows (usually 7 + 7) (Fig. 505).

    Brachythemis leucosticta

    (*B. fuscopalliata* is still undescribed)

–   Labial palpus with 6 or more setae; 18 or more premental setae   6

---

* *Sympetrum fonscolombei* (Fig. 501) may also key out here. It may be distinguished, at least from *Crocothemis erythraea* (Fig. 500), which has 10–11 setae on the labial palpus, by having 12–13 setae on its labial palpus.

6. Labial palpus with 6–7 setae; 10–14 long, and 6–10 short premental setae (Figs. 505–506). **Trithemis**
–   Labial palpus with 8 or more setae; premental setae otherwise arranged   7
7. Eight palpal setae and about 20 premental setae (Figs. 508–509). **Zygonyx torrida**
–   More than 9 palpal setae   8
8. Articulation between prementum and mentum, when mask at rest, situated between second and third pair of legs; examples of prementum structures: Figs. 501–502.

    Sympetrum
–   Articulation between prementum and mentum, when mask at rest, situated at level of second pair of legs. Habitus and various structures: Figs. 510–511.

    Selysiothemis nigra

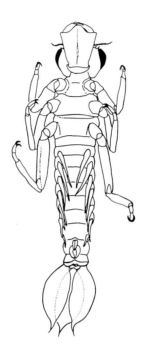

Fig. 446: *Epallage fatime* (Charpentier, 1840), female, ventral view
(after Schneider 1984)

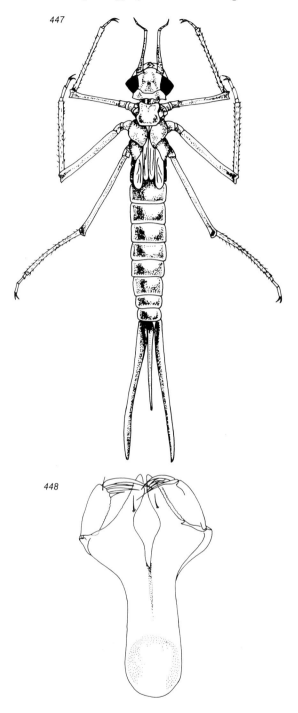

Figs. 447–448: *Calopteryx syriaca* Rambur, 1842
447. dorsal view; 448. pronotum

Fig. 449: *Sympecma fusca* (Vander Linden, 1820), prementum
(after Conci & Nielsen, 1956)

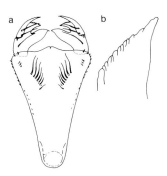

Fig. 450: *Lestes viridis parvidens* Artobolevski, 1929
a. prementum, b. distal ending of labial palp, enlarged
(after Conci & Nielsen, 1956)

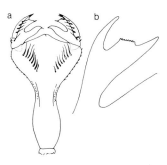

Fig. 451: *Lestes macrostigma* (Eversmann, 1836)
a. prementum; b. labial palp
(after Conci & Nielsen, 1956)

Fig. 452: *Lestes barbarus* (Fabricius, 1798), prementum
(after Conci & Nielsen, 1956)

Fig. 453: *Lestes virens* (Charpentier, 1825), prementum
(after Conci & Nielsen, 1956)

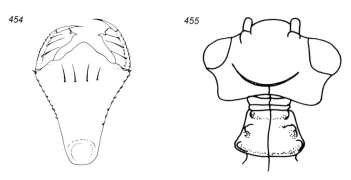

Figs. 454–455: *Platycnemis* sp.
454. prementum;
455. head and pronotum
(note lateral bosses as in imago)

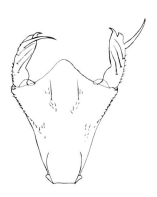

Fig. 456: *Pseudagrion* sp., prementum
(after Chutter, 1961)

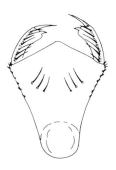

Fig. 457: *Ceriagrion* sp., prementum

262

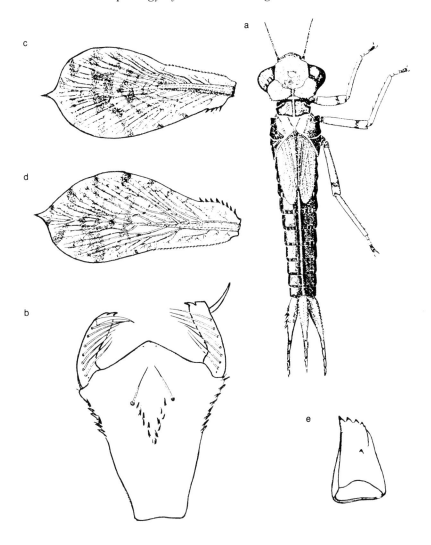

Fig. 458: *Ceriagrion glabrum* (Burmeister, 1839)
a. habitus; b. prementum; c–d. caudal lamellae; e. mandibula
(after Corbet, 1957)

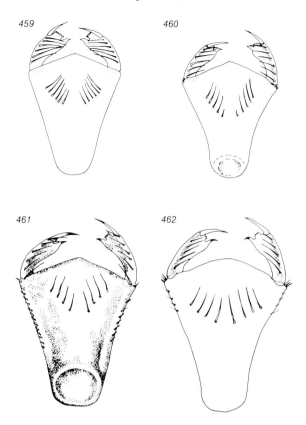

Figs. 459–462: Prementum of various coenagrionids
459. *Coenagrion scitulum* (Rambur, 1842);
460. *C. puella* (Linnaeus);
461. *C. lindeni* (Selys)
462. *Erythromma viridulum* (Charpentier);
(after Conci & Nielsen, 1956)

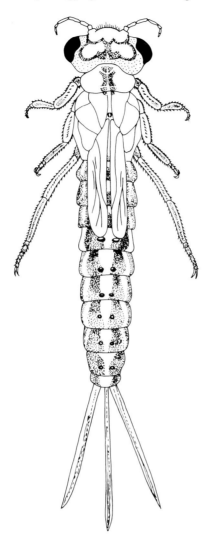

Fig. 463: *Ischnura fountainei* Morton, 1905; dorsal view

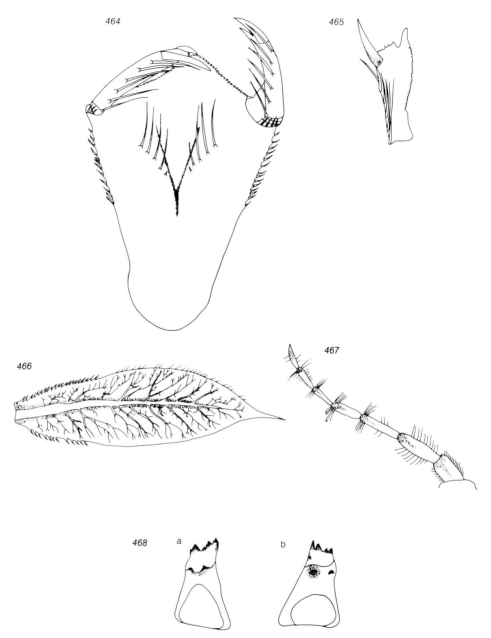

Figs. 464–468: *Ischnura fountainei* Morton, 1905
464. prementum; 465. labial palps;
466. median caudal lamella; 467. antenna;
468. a–b. left and right mandibules

469

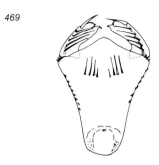

Fig. 469: *Ischnura elegans*, prementum
(after Conci & Nielsen, 1956)

470

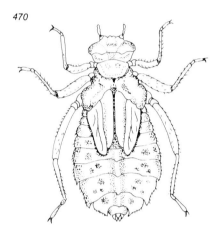

471

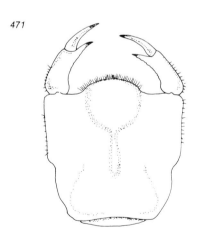

Figs. 470–471: *Lindenia tetraphylla* (Vander Linden, 1825)
470. dorsal view; 471. prementum
(after Conci & Nielsen, 1956)

267

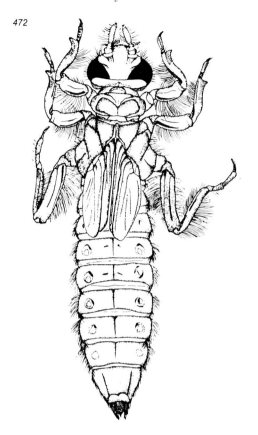

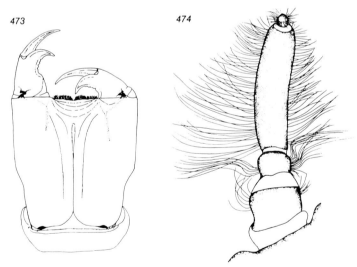

Figs. 472–474: *Gomphus davidi* Selys, 1887
472. dorsal view; 473. prementum; 474. antenna
(after Schneider, 1983a)

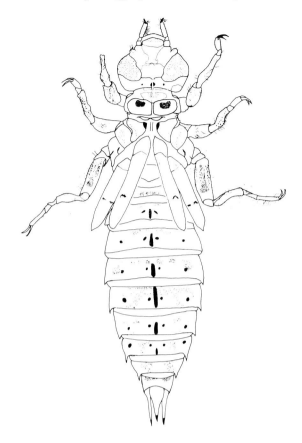

Fig. 475: *Paragomphus sinaiticus* (Morton, 1929), dorsal view
(after Martens & Dumont, 1983)

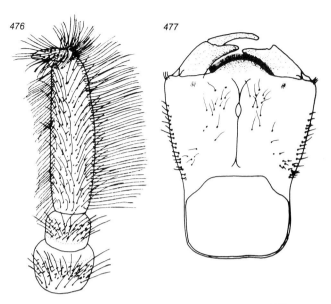

Figs. 476–477: *Paragomphus sinaiticus* (Morton, 1929)
476. antenna; 477. prementum
(after Martens & Dumont, 1983)

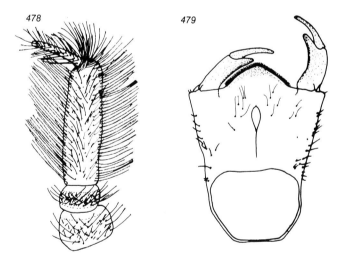

Figs. 478–479: *Paragomphus genei* (Selys, 1841)
478. antenna; 479. prementum
(after Martens & Dumont, 1983)

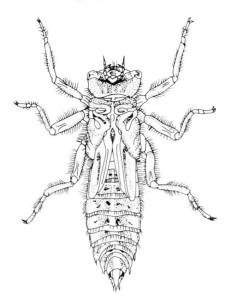

Fig. 480: *Cordulegaster* sp., dorsal view

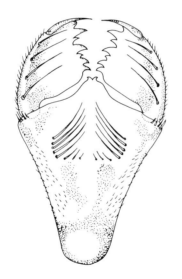

Fig. 481: *Cordulegaster* sp., prementum
(after Conci & Nielsen, 1956)

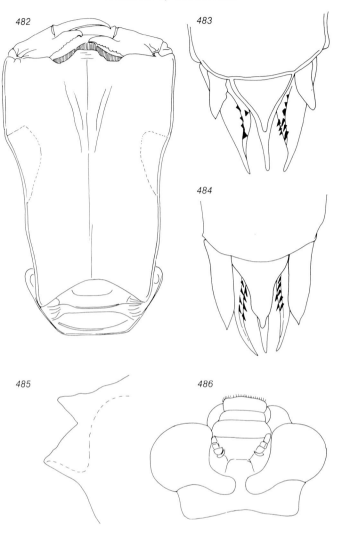

Figs. 482–486: *Caliaeschna microstigma* (Schneider, 1845)
482. prementum; 483–484. anal pyramid in male and female
(note pronounced sexual dimorphism);
485. precoxal armature; 486. head, dorsal view

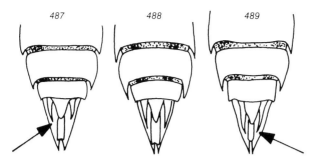

Figs. 487–489: Anal pyramids in aeschnids
487. *Aeshna affinis* Vander Linden, 1820;
488. *A. mixta* Latreille, 1805; 489. *Anaciaeschna isoceles*
(arrows point to tips of paraprocts)
(after Conci & Nielsen, 1956)

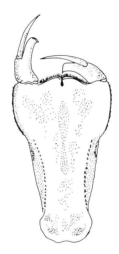

Fig. 490: *Anax imperator* Leach, 1815; prementum
(after Comte-Sart, 1965)

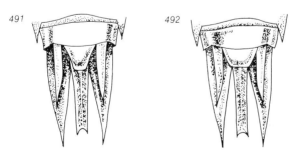

Figs. 491–492: *Anax* sp., anal pyramid
491. *A. imperator* Leach, 1815;
492. *A. parthenope* (Selys, 1839)

273

493

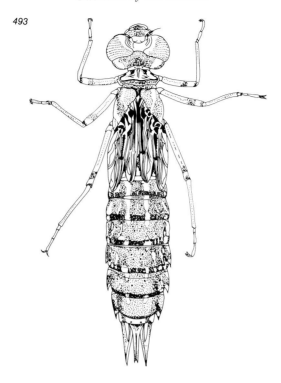

494

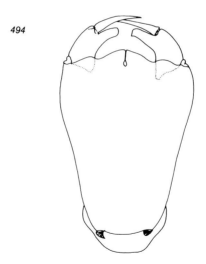

Figs. 493–494: *Hemianax ephippiger* (Burmeister, 1839)
493. habitus; 494: prementum
(after Degrange & Seassau, 1970)

274

Fig. 495: *Aeshna affinis* Vander Linden, 1820; prementum
(after Conci & Nielsen, 1956)

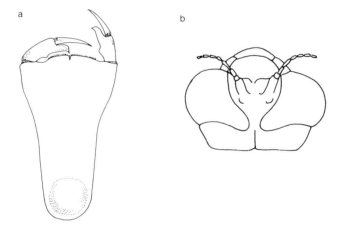

Fig. 496: *Aeshna mixta* Latreille, 1805; a. prementum
(after Conci & Nielsen, 1956)
b. head

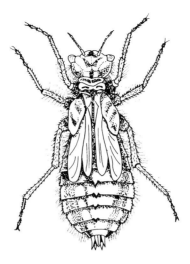

Fig. 497: *Libellula* sp., habitus
(after Conci & Nielsen, 1956)

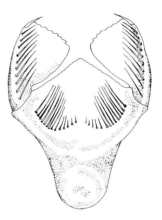

Fig. 498: *Libellula depressa* Linnaeus, 1758; prementum
(after Conci & Nielsen, 1956)

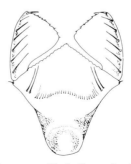

Fig. 499: *Orthetrum brunneum* (B. de Fonscolombe, 1837); prementum
(after Conci & Nielsen, 1956)

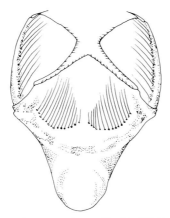

Fig. 500: *Crocothemis erythraea* (Brullé, 1832); prementum
(after Conci & Nielsen, 1956)

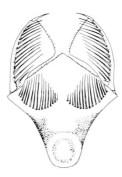

Fig. 501: *Sympetrum fonscolombei* (Selys, 1837); prementum
(after Conci & Nielsen, 1956)

Fig. 502: *Sympetrum pedemontanum* (Allioni, 1766); prementum
(after Conci & Nielsen, 1956)

277

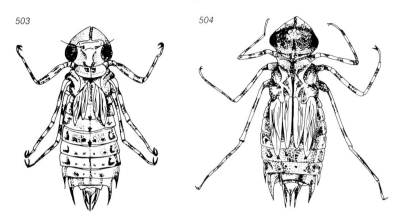

Figs. 503–504: *Pantala flavescens* (Fabricius, 1798)
503. immature larva (note length of wing stubs); 504. fully grown larva

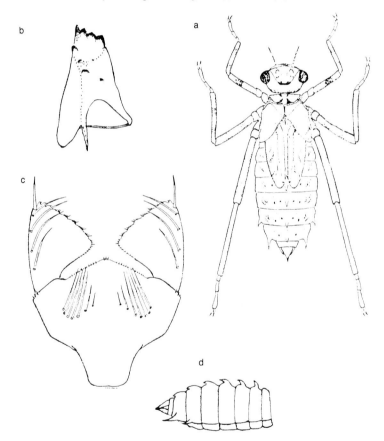

Fig. 505: *Brachythemis leucosticta* (Burmeister, 1839)
a. habitus; b. mandible; c. prementum;
d. lateral view of abdomen
(after Corbet, 1956)

278

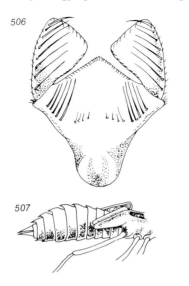

Figs. 506–507: *Trithemis annulata* (Palisot de Beauvois, 1805)
506. prementum; 507. abdomen, lateral view
(after Conci & Nielsen, 1956)

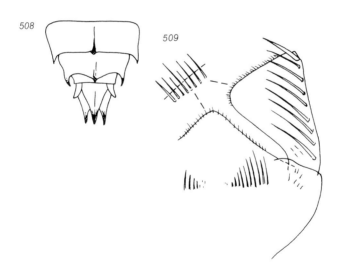

Figs. 508–509: *Zygonyx torrida* (Kirby, 1889)
508. anal pyramid; 509. prementum
(after Barnard, 1937)

279

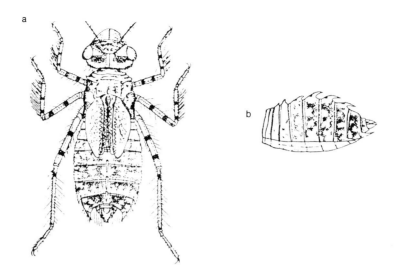

Fig. 510: *Selysiothemis nigra* (Vander Linden, 1825)
a. habitus; b. abdomen, lateral view

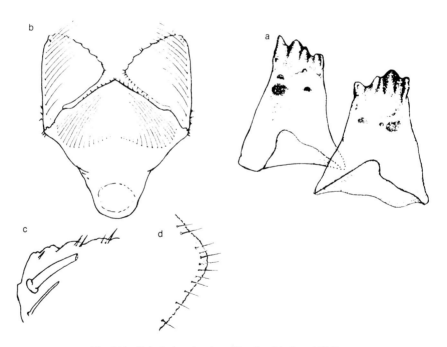

Fig. 511: *Selysiothemis nigra* (Vander Linden, 1825)
a. mandibles; b. prementum; c. insertion of movable hook;
d. distal margin of median lobe
(after Comte-Sart, 1960)

REFERENCES

Adetunji, J.F. & M.J. Parr (1974) 'Colour change and maturation in *Brachythemis leucosticta* (Burmeister)', *Odonatologica*, 3:13–20.

Aguesse, P.A. (1968) *Les Odonates de l'Europe Occidentale, du Nord de l'Afrique et des Iles Atlantiques*, Masson et Cie, Paris, 258 pp.

Akramowski, N.N. (1948) 'The dragonfly fauna of the Soviet Armenia', *Zool. Zb. Akad. NAUK Armenian S.S.R., Erevan*, 5:117–188 (in Russian).

Allioni, C., (1766) 'Manipulus insectorum taurinensium', *Méletemata Phil. Mathém. Soc. r. Turin* (1762–65), 3, 7:185–198.

Andres, A. (1928) 'The Dragonflies of Egypt', *Mém. Soc. r. ent. Égypte*, 3:7–43.

— (1929) 'Note on the Egyptian dragonflies', *Bull. Soc. r. ent. Égypte*, N.S., 13:9.

Artobolevski, G. (1929a) 'Les Odonates de la Crimée', *Bull. Soc. Nat. Crimée*, Sinferopol, 11:139–150 (in Russian).

— (1929b) 'Les Odonates du Daghestan', *Rev. russe. Ent.*, 23:225–240 (in Russian).

Asahina, S. (1973) 'The Odonata of Iraq', *Jap. J. Zool.*, 17:17–36.

— (1974) 'An additional note to the Odonata of Iraq', *Kontyû*, Tokyo, 42:107–109.

Barnard, K.H. (1937) 'Notes on dragon-flies (Odonata) of the S.W. Cape, with descriptions of the nymphs, and of new species', *Ann. S. Afr. Mus.*, 32:169–260.

Barraud, P.J. (1920) 'Entomology in the Holy Land', *Entomologist*, 53:170–175.

— (1923) 'Entomology in the Holy Land', *Entomologist*, 56:58–62.

Bartenef, A.N. (1912) 'Die palaearktischen und ostasiatischen Arten und Unterarten der Gattung *Calopteryx* Leach', *Lab. Arb. Zool. Kabin. Univ. Warschau*, 1, 1911 (1912) (sép. 1–193 + erratum) (in Russian).

— (1925) 'Contribution à l'Odonatofaune des monts de la Caucasie', *Bull. Mus. Géorgie*, 2:28–86 (in Russian).

— (1929) 'Neue Arten und Varietaten der Odonaten des West-Kaukasus', *Zool. Anz.*, 85:54–68.

— (1930) 'Uber *Calopteryx splendens* und ihre Biotypen besonders die Westasiatischen', *Zool. Jb. Syst.*, 58:521–540.

Beauvois, Palisot de (1805) *Insectes recueillis en Afrique et en Amérique*, Paris, xvi + 276 pp.

Bodenheimer, F.S. (1935) *Animal Life in Palestine*, L. Mayer, Jerusalem, 507 pp.

— (1937) 'Prodromus Faunae Palestinae', *Mém. Inst. Égypte*, 33:230–231 (Odonata).

— (1938) 'On the presence of an Irano-Turanian relict fauna in North Africa', *Mém. Soc. Biogéogr.*, 6:67–79.

Bolivar, I. (1893) 'Liste des Orthoptères recueillis en Syrie par le Dr. Theod. Barrois', *Rev. biol. nord Fr.*, 5:476–489.

Brauer, F. (1865) 'Bericht über die von Hern Baron Ransonnet am rothen Meere und auf Ceylon gesammelten Neuropteren', *Verh. zool.-bot. Ges. Wien*, 15:1009–1018.

— (1866) Neuropteren, in: *Reise der Osterreichischen Fregatte Novara um die Erde in den Jahren 1857, 1858, 1859 unter den Befehlen des Commodore B. Von Wüllerstorf-Urbair*, Zool. Teil., Wien, 107 pp.

— (1868) 'Verzeichniss der bis jetzt bekannenten Neuropteren im Sinne Linné's', *Verh. zool.-bot.Ges. Wien*, 18:359–416, 711–742.

Brullé, A. (1832) *Expedition scientifique de Morée. 3,1 Partie. Zoologique. 2. Section. Des animaux articulés*, Paris, Levrault, pp. 1–29, 64–395 (Odonata: 99:101–106) (Atlas: 1835, Odonata, Plate 32, figs. 4–9).

Buchholtz, C. (1955) 'Eine vergleichende Ethologie der Orientalische Calopterygiden (Odonaten), als Beitrag zu ihren systematischen Deutung', *Z. Tierpsychol.*, 12:364–386.

— (1956) 'Eine Analyse des Paarungsverhaltens und der dabei wirkenden Auslöser bei den Libellen *Platycnemis pennipes* Pall. und *Pl. dealbata* Klug', *Z. Tierpsychol.*, 13:17–25.

Buchholz, K.F. (1959) 'Odonaten aus dem Ennedigebirge (Fr. Equat. Afr.). nebst Bemerkungen über einige aethiopische Arten', *Bonn. zool. Beitr.*, 1:75–98.

Burmeister, F. (1839) *Handbuch der Entomologie*, Vol. 2:805–852, 1016–1017, Eslin, Berlin.

Cabot, L. (1889) 'The immature state of the Odonata. Part III. Subfamily Cordulina', *Mem. Mus. comp. Zool. Harv.*, 17:1–52.

Calvert, P.P. (1892) 'Preliminary notes on some African Odonata', *Trans. Am. ent. Soc.*, 19:161–164.

Carchini, G. (1983) 'A key to the Italian Odonata Larvae', *S.IO Rapid Communications*, Utrecht, 1, 101 pp.

Carfi, S. (1974) 'Contribution to the knowledge of Somalian Odonata', *Monitore zool. ital.*, Suppl., N.S., 5:147–181 (contains a review on the Odonata of Somalia, with complete bibliography for the area of the Horn of Africa).

Charpentier, T. de (1825) *Horae entomologicae*, Gosohorsky, Wratislaviae, pp. XVI + 259 (Odonata XII, 1–50).

— (1840) *Libellulinae europaea, descriptae ac depictae*, Leopold Voss, Lipsiae, 180 pp.

Chutter, F.M. (1961) 'Certain aspects of the morphology and ecology of several species of *Pseudagrion* Selys (Odonata)', *Arch. Hydrobiol.*, 57:430–463.

Comstock, J.H. & J.G. Needham (1898) 'The wings of insects', *Am. Nat.*, 32:43–48, 81–89, 231–257, 335–340, 413–424, 561–565, 769–777, 903–911.

Comte-Sart, A. (1960) 'Biografia de la *"Selysiothemis nigra"* Vd. L. (Odonatos)', *Graellsia*, 18:73–115.

— (1965) 'Distribución, ecologia y biocenosis de los Odonatos ibericos', *Publnes Inst. Biol. apl., Barcelona*, 39:33–64.

Conci, C. & C. Nielsen (1956) 'Odonata', in: *Fauna d'Italia*, Calderini, Bologna, Vol. 1, 308 pp.

Corbet, P.S. (1953) 'A terminology for the labium of larval Odonata', *Entomologist*, 86:191–196.

— (1956) 'Larvae of East African Odonata. 2–3', *Entomologist*, 89:148–151.

— (1957) 'Larvae of East African Odonata 6–8', *Entomologist*, 90:28–34.

— (1962) *A Biology of Dragonflies*, Witherby, London, 247 pp.

— (1980) 'Biology of Odonata', *A. Rev. Ent.*, 25:189–217.

Cowley, J. (1934a) 'Changes in the generic names of the Odonata', *Entomologist*, 67:200–205.

— (1934b) 'Notes on some generic names of Odonata', *Entomologist's mon. Mag.*, 70:240–247.

— (1940) 'A list of the Odonata of the islands of the western mediterranean area', *Proc. R. ent. Soc. Lond.*, B, 9:172–178.

— (1944) 'Additions to the list of Odonata of the eastern mediterranean islands', *Proc. R. ent. Soc. Lond*, B, 13:88–89.

De Marmels, J. (1975) 'Die Larve von *Hemianax ephippiger* (Burmeister, 1839)', *Odonatologica*, 4:259–263.

Desjardins, J. (1835) '*Libellula limbata, L. semihyalina, L. bimaculata* de Maurice', *Ann. Soc.ent. Fr.* 4(1):1, 3–4.

282

References

De Villers, C. (1789) *Caroli Linnaei Entomologia*. Lugduni, Piestre & Delamollière, 657 pp. (Odonata: 1–15).

Dopffer, J. (1912) 'Notes sur les Libellules', *Bull. Soc. ent. Égypte*, 4:124–128.

Drury, D. (1770) *Illustrations of natural history, wherein are exhibited upwards of two hundred and forty figures of exotic insects*, White, London, 1:130 pp.

Dumont, H.J. (1967) 'A possible scheme of the migration of *Crocothemis erythraea* (Brulle) — populations from the Camargue (Odonata: Libellulidae)', *Biol. Jrb. Dodonaea*, pp. 222–227.

— (1972) 'Occurrence of *Brachythemis fuscopalliata* (Selys, 1887) in the East Mediterranean area (Anisoptera: Libellulidae)', *Odonatologica*, 1:241–244.

— (1973) 'The genus *Pseudagrion* Selys in Israel and Egypt, with a key to the regional species (Insecta:Odonata)', *Israel J. Zool.*, 22:169–195 (published 1974).

— (1974) '*Agriocnemis sania* Nielsen, 1959 (Odonata: Zygoptera) from Israel and Sinai, with a redescription of the species and distributional and ecological notes', *Israel J. Zool.*, 23:125–134 (published 1975).

— (1975) 'Endemic dragonflies of late Pleistocene age of the Hula Lake area (northern Israel), with notes on the Calopterygidae of the Rivers Jordan (Israel, Jordan) and Litani (The Lebanon) and description of *Urothemis edwardsi hulae* subspec. nov. (Libellulidae)', *Odonatologica*, 4(1):1–9.

— (1977a) 'A survey of the dragonfly fauna of Tunisia', *Bull. Annls Soc. r. ent. Belg.*, 113:63–94.

— (1977b) 'A review of the dragonfly fauna of Turkey and adjacent mediterranean islands (Insecta:Odonata)', *Bull. Annls Soc. r. ent. Belg.*, 113:119–171.

— (1977c) 'Sur une collection d'Odonates de Yougoslavie, avec notes sur la faune des territoires adjacents de Roumanie et de Bulgarie', *Bull. Annls Soc. r. ent. Belg.*, 113:187–209.

— (1977d) '*Orthetrum abbotti* Calvert, 1892, a new Ethiopian representative in the Palaearctic fauna (Anisoptera: Libellulidae)', *Odonatologica*, 6(3):199–203.

— (1978a) 'On confusion about the identity of *Pseudagrion acaciae* Förster, 1906, with the description of *Pseudagrion niloticum* n. sp., and on the identity of *Pseudagrion hamoni* Fraser, 1955 (Zygoptera: Coenagrionidae)', *Odonatologica*, 7(2):123–133.

— (1978b) 'Odonates d'Algérie, principalement du Hoggar et d'Oasis du Sud', *Bull. Annls Soc. r. ent. Belg.*, 114:99–106.

— (1978c) 'Odonata from Niger, with special reference to the Aïr mountains', *Revue zool. afr.*, 92:303–316.

— (1979) 'Limnologie van Sahara en Sahel', Thesis, University of Gent, 557 pp.

— (1980) 'The dragonfly fauna of Egypt and the rôle of the Nile in its origin and composition', *Water Supply & Management*, 4:29–34.

Dumont, H.J. & S. Dumont (1969) 'A biometrical analysis of the dragonfly *Ischnura elegans elegans* (Vander Linden) with special reference to its chloride-tolerance and generation number', *Biol. Jrb. Dodonaea*, pp. 50–60.

Dumont, H.J. & K. Martens (1984) 'Dragonflies (Insecta, Odonata) from the Red Sea Hills and the main Nile in Sudan', *Hydrobiologia*, 110:181–190.

Dumont, H.J. & W. Schneider (1984) 'On the presence of *Cordulegaster mzymtae* Barteneff, 1929 in Turkey, with a discussion of its geographic distribution and taxonomic position (Anisoptera: Cordulegastridae)', *Odonatologica*, 13(3):467–476.

El Rayah, E.A. & F.T. El Din Abu Shama (1978) 'Notes on morphology and bionomy of the dragonfly *Trithemis annulata scortecii* Nielsen (Odonata:Anisoptera), as a predator on mosquito larvae', *Z. angew. Ent.*, 85:81–86.

283

Eversmann, E. (1836) 'Libellulinae, Wolgam fluvium inter et montes Uralensis observatae, & Libellulinarum species novae, quas inter Wolgam fluvium et montes Uralensis observavit Dr. Eduard Eversmann', *Bull. Soc. imp. Moscou*, 9:233, 235–248.

Fabricius, J.C. (1775) *Systema Entomologiae*, Kortuis, Flensburg & Lipsia, 832 pp. (Odonata, pp. 420–426).

— (1793) *Entomologia systematica emendata et aucta*, Hafniae, Vol. 2, viii + 520 pp. (Odonata V, pp. 373–388).

— (1798) *Supplementum entomologiae systematicae*, Proft & Storch, Hafniae, 2 + 572 pp. (Odonata, pp. 283–287).

Fonscolombe, M. Boyer de (1837) 'Monographie des Libellulines des environs d'Aix', *Annls Soc. ent. Fr.*, 6:129–150.

— (1838) 'Monographie des Libellulines des environs d'Aix', *Annls Soc. ent. Fr.*, 7:75–106, 547–575.

Förster, F. (1898) 'Odonaten des Transvaalstaates', *Ent. Nachr.*, 24:166–172.

— (1906a) 'Die Libellulidengattungen von Afrika und Madagascar', *Jber. Ver. Naturk. Mannheim*, 71–72:1–71.

— (1906b) 'Libellen der Foschungsreise durch Südschoa, Galla und die Somaliländer von Carlo Freiherr von Erlanger', *Jahrb. Nassaue Ver. Naturk.*, 59:301–344.

— (1909) 'Odonata, in: Kneucker, F., Zoologische Ergebnisse zweies in den Jahren 1902 und 1904 durch die Sinai Halbinsel unternommener botanischer Studiereisen, nebst zoologischen Beobachtungen aus Aegypten, Palästina und Syrien', *Verh. naturw. Ver. Karlsruhe*, 21:1–90 (Odonata p. 44).

Fraser, F.C. (1917) 'The female of the dragonfly *Brachythemis fuscopalliata* (Ris)', *J. Bombay nat. Hist. Soc.*, 25:282–283.

— (1929) 'A revision of the Fissilabioidea (Cordulegasteridae, Petaliidae, and Petaluridae) (Order Odonata). Part I. Cordulegasteridae', *Mem. Indian Mus.*, 9(3):69–167.

— (1933) Odonata, Vol. I, in: *The Fauna of British India*, Taylor & Francis, London, 423 pp.

— (1934) Odonata, Vol. II, in: *The Fauna of British India*, Taylor & Francis, London, 398 pp.

— (1936) Odonata, Vol. III, in: *The Fauna of British India*, Taylor & Francis, London, 461 pp.

— (1957) 'A reclassification of the order Odonata', *Proc. zool. Soc. N.S.W.*, 1957:1–133.

Gadeau de Kerville, H. (1926) Odonates, in: *Voyage Zoologique d'Henri Gadeau de Kerville en Syrie (Avril-Juin 1908)*, Baillére, Paris, pp. 78–80.

Gambles, R.M. (1960) 'Seasonal distribution and longevity in Nigerian dragonflies', *Jl W. Afr. Sci. Ass.*, 6:18–26.

Hagen, H.A. (1856) 'Die Odonaten des russischen Reiches', *Stettin. ent. Ztg*, 70:363–381.

— (1861) *Synopsis of the Neuroptera of North America, with a list of the South American species*, Smithsonian Institution, Washington, 347 pp.

— (1863) 'Die Odonaten- und Neuropteren-Fauna Syriens und Kleinasiens', *Wien. ent. Monatsschr.*, 7:193–199.

— (1867) 'Die Neuropteren der Insel Cuba', *Stettin. ent. Ztg*, 28:215–232.

Hammond, C.D. (1977) *The Dragonflies of Great-Britain and Ireland*, Curwen Press, London, 115 pp. (Key to larvae by A. E. Gardner).

Happold, D.C.D. (1966) 'Dragonflies of Jebel Marra, Sudan', *Proc. R. ent. Soc. Lond.*, A, 41:87–91.

Hariri, G. El- (1968) *A List of Recorded Syrian Insects and Acari*, Al Chark-Lahlouh, Aleppo, 157 pp. (Odonata pp. 18–18).

Heymer, A. (1973) 'Verhaltensstudien an Prachtlibellen', *J. comp. Ethol.*, Suppl. 11:1–100.

— (1975) 'Der stammesgeschichtliche Aussagewert des Verhaltens der Libelle *Epallage fatime* Charp. 1840', *Z. Tierpsychol.*, 37:163–181.

References

Karsch, F. (1890) 'Uber Gomphiden', *Ent. Nachr.*, 16:370–382.

— (1893) 'Insekten der Berglandschaft Adeli im Hinterlande von Togo, Westafrika', *Berlin ent. Z.*, 38:17–28.

Kempny, P. (1908) 'Beitrag zur Neuropterenfauna des Orients', *Verh. zool.-bot. Ges. Wien*, 58:263–270.

Kimmins, D.E. (1934) 'Report on the insects collected by Colonel R. Meinertzhagen in the Ahaggar mountains. III. Odonata', *Ann. Mag. nat. Hist.*, Ser. 10, 74:173–175.

— (1950) 'Results of the Armstrong College Expedition to Siwa Oasis (Libyan Desert) 1935 under the leadership of Prof. J. Omer-Cooper', *Bull. Soc. Fouad 1er Ent.*, 34:151–154 (Odonata).

Kirby, W.F. (1886) 'On a small collection of Dragonflies from Murree and Campbellpore (N.W. India, received from Major J.W. Yerbury, R.A.', *Proc. zool. Soc. Lond.*, 1886:325–329.

— (1889) 'A revision of the subfamily Libellulinae, with descriptions of new genera and species', *Trans. zool. Soc. Lond.*, 12:249–348.

— (1890) *A Synonymic Catalogue of Neuroptera Odonata or Dragonflies with an appendix on fossil species*, Gurney & Jackson, London, ix + 202 pp.

Kumar, A. (1971) 'The larval stages of *Orthetrum brunneum brunneum* (Fonscolombe) with a description of the last instar larvae of *Orthetrum taeniolatum* (Schneider)', *J. nat. Hist.*, 5:121–132.

Lacroix, J.-L. (1924) 'Sur quelques Odonates d'Afrique de la collection du Museum', *Bull. Mus. natn. Hist. nat., Paris*, 80:215–222.

Laidlaw, F.F. (1913) 'Note on the dragonflies of Syria and the Jordan valley', *J. Asiat. Soc. Beng.*, 1913:219–220.

Latreille, P.A. (1805) *Histoire naturelle, générale et particulière, des Crustacés des Insectes*, Dufart, Paris, 432 pp. (Odonata, pp. 5–16).

Leach, W.E. (1815) Entomology, in: Brewster, *Edinburgh Encyclopaedia*, William Blackwood, Edinburgh, Vol. 9, pp. 57–172 (Odonata: pp. 136–137).

Le Roi, O. (1915) 'Odonaten aus der Algerischen Sahara von der Reise von Freiherrn H. Geyr von Schweppenburg. Mit einer Übersicht der Nordafrikanisher Odonaten-fauna', *Dt. ent. Z.*, pp. 609–634.

Lieftinck, M.A. (1966) 'A survey of the dragonfly fauna of Morocco (Odonata)', *Bull. Inst. r. Sci. nat. Belg.*, 42(35):1–63

Linnaeus, C. (1758) *Systema Naturae. Regnum Animale*, 10th Edition, Stockholm, 824 pp. (Odonata: pp. 543–546).

Lohmann, H. (1981) 'Zur Taxonomie einiger *Crocothemis* — Arten, nebst Beschreibung einer neuen Art von Madagaskar (Anisoptera: Libellulidae)', *Odonatologica*, 10:109–116.

Longfield, C. (1932a) 'List of Odonata from Asia Minor collected by Mr. B. P. Uvarov (July-August 1931)', *Bol. Soc. esp. Hist. nat. Madrid*, 32:159–160.

— (1932b) 'A new species of the genus *Urothemis* from Southern Arabia, and some comments on the species of Odonata inhabiting the Qara mountains', *Stylops*, 1:34–35.

— (1955) 'The Odonata of N. Angola. Part I. A revision of the African species of the genus *Orthetrum*', *Publções cult. Co. Diam. Angola*, 27:11–64.

McLachlan, R. (1899) 'Remarques sur quelques Odonates de l'Asie Mineure méridionale, comprenant une espèce nouvelle pour la faune paléarctique', *Annls Soc. ent. Belg.*, 43:301–302.

Martens, K. & H.J. Dumont (1983) 'Description of the larval stages of the desert dragonfly *Paragomphus sinaiticus* (Morton), with notes on the larval habitat, and a comparison with three related species (Anisoptera:Gomphidae)', *Odonatologica*, 12(3):285–296.

Martin, R. (1894) 'Odonates de Chypre', *Bull. Soc. zool. Fr.*, 19:135–138.

— (1908–1909) Aeschnines, in: *Cat. Colln. Zool. Baron E. de Selys Longchamps*, 18–20:1–223.

— (1909) 'Notes sur trois Odonates de Syrie', *Bull. Soc. ent. Fr.*, 12:212–214.

— (1912) 'Les Odonates', *Ann. Hist. nat. Délég. Perse*, 2(1):5–9.

— (1926) 'Notes sur trois Odonates de Syrie (Névropterès)', in: *Voyage zoologique d'Henri Gadeau de Kerville en Syrie (Avril–Juin 1908)*, Baillère, Paris, pp. 275–278 (a reprint of 1909).

May, E. (1933) Libellen oder Wasserjungfern (Odonata), in: *Die Tierwelt Deutschlands*, G. Fisher, Jena, Vol. 27, 124 pp.

Meyer-Dür, L. (1874) 'Die Neuropteren-Fauna der Schweiz, bis auf heutige Erfahrung zusammengestellt', *Mitt. schweiz ent. Ges.*, 4:281–436.

Morton, K.J. (1905) 'Odonata collected by Miss Margaret E. Fountaine in Algeria, with description of a new species of *Ischnura*', *Entomologist's mon. Mag.*, 41:145–149.

— (1915a) 'Some palaearctic species of *Cordulegaster*', *Trans. R. ent. Soc. Lond.*, 1915:272–290.

— (1915b) 'Notes on Odonata from the environs of Constantinople', *Entomologist*, 48:129–134.

— (1919) 'Odonata from Mesopotamia', *Entomologist's mon Mag.*, 3:143–151; 183–196.

— (1920a) 'Odonata collected in North-Western Persia by captain P.A. Buxton', *Entomologist's mon. Mag.*, 56:82–87.

— (1920b) 'Odonata collected in Mesopotamia by the late Major R. Brewitt-Taylor', *Ann. Mag. nat. Hist.*, S.9, 5:293–303.

— (1921) 'Neuroptera, Mecoptera, and Odonata from Mesopotamia and Persia', *Entomologist's mon. Mag.*, 57:213–225.

— (1922) 'Further notes on the Odonata of Constantinople and adjacent parts of Asia Minor', *Entomologist*, 55:80–82.

— (1924) 'The dragonflies of Palestine, based primarily on collections made by Dr. P. A. Buxton, with notes on the species of adjacent regions', *Trans. R. ent. Soc. Lond.*, 1924:25–44.

— (1929) 'Odonata from the Sinai peninsula, Suez and Palestine including a new species of *Mesogomphus*', *Entomologist's mon. Mag.*, 65:60–63.

Navas, L. (1909) 'Neuropteros de Egypto', *Broteria*, 8:4.

— (1911) 'Algunos ortopteros y Neuropteros de Palestina', *Rev. Montserratina*, 1911:1–4.

— (1929 [1930]) 'Insectos de la Cirenaica', *Revta Acad. Cienc. Zaragoza*, 13:13–28 (Odonata pp. 13–14).

— (1931) 'Spedizione scientifica all oasi di Cufra. Insetti Neuroteri ed affini', *Annali Mus. civ. Stor. nat. Giacomo Doria, Genova*, 55:409–421 (Odonata pp. 409–410).

— (1932) 'De las cazas del Sr. Gadeau de Kerville en el Asia Menor', Ve Congr. int. Ent., 1932:221–225 (Odonata pp. 221–222).

— (1936) 'Mission au Tibesti: Paranévropterès et Névropterès', *Mém. Acad. Sci. Paris*, 2, 62:72–74.

Newman, E. (1833) 'Entomological notes', *Entomologist's mon. Mag.*, 1:505–514.

Nielsen, C. (1935) 'Odonati del Fezzan raccolti del Prof. G. Scortecci e catalago delle specie finora catturate', *Atti Soc. ital. Sci. nat.*, 74:372–382.

— (1959) 'Una nuova specie del genere *Agriocnemis* Selys (Odonata) di Gat (Fezzan)', *Riv. Biol. colon.*, 16 (1956–1958):33–40.

Paulson, D.R. (1974) 'Reproductive isolation in damselflies', *Syst. Zool.*, 23:40–49.

Pfau, H.K. (1970) 'Die Vesica Spermalis von *Aeschna cyanea* Müll. und *Cordulegaster annulatus* Latr., ihre Anatomie, Funktion und phytogenetische Bedeutung', *Tombo*, 13:5–11.

— (1971) 'Struktur und Funktion des sekundären Kopulations-apparates der Odonaten (Insecta, Palaeoptera), ihre Wandlung in der Stammesgeschichte und Bedeutung für die adaptive Entfaltung der ordnung', *Z. Morph. Ökol. Tiere*, 70:281–371.

References

Pinhey, E.C.G. (1951) 'The dragonflies of Southern Africa', *Transv. Mus. Mem.*, 5:335 pp.

— (1961) *A Survey of the Dragonflies of Eastern Africa*, British Museum, London, 214 pp.

— (1962) 'A descriptive catalogue of the Odonata of the African continent (up to December 1959)', *Publções cult. Co. Diam. Angola*, 59:322 pp..

— (1964) 'A revision of the African members of the genus *Pseudagrion* Selys (Odonata)', *Revta Ent. Moçamb.*, 7:5–196.

— (1970a) 'A new approach to African *Orthetrum* (Odonata)', *Occ. Pap. natn. Mus. Rhodesia*, 4, 30b:261–321.

— (1970b) 'Monographic study of the genus *Trithemis* Brauer (Odonata:Libellulidae)', *Mem. ent. Soc. sth. Afr.*, 11:159 pp.

— (1974) 'A revision of the African *Agriocnemis* Selys and *Mortonagrion* Fraser (Odonata: Coenagrionidae)', *Occ. Pap. natn. Mus. Monum. Rhod.*, B. 5:171–278.

— (1980) 'A revision of African Lestidae (Odonata)', *Occ. Pap. natn. Mus. Monum. Rhod.*, S.B, 6:329–479.

Por, F.D. (1975) 'An outline of the Zoogeography of the Levant', *Zoologica Scripta*, 4:5–20.

Rambur, J.P. (1842) *Histoire naturelle des insectes. Neuroptères*, Roret, Paris. pp. 17 + 534.

Ris, F. (1897) 'Note sur quelques Odonates de l'Asie Centrale', *Annls Soc. ent. Belg.*, 41:42–50.

— (1909) 'Abessinische Libellen, gesammelt von Dr. Eduard Rüppell', *Ber. senckenb. naturf. Ges.*, 40:21–27.

— (1909–1919) *Libellulinen monographisch bearbeitet*, Cat. Collns. Zool. Baron E. de Selys Longchamps, 9–16, pp. 1–1278.

— (1911) 'Libellen von Tripolis und Barka', *Zool. Jb. Syst.*, 30:643–650.

— (1912) 'Ergebnisse der mit Subvention aus der Erbschaft Treitl unternommenen Zoologischen Forschungsreise Dr. Franz Werner's nach dem Aegyptischen Sudan und Nord-Uganda. 17 Libellen', *Sber. Akad. Wiss. Wien*, 121:149–170.

— (1913) 'Expedition to the central-western Sahara by Ernst Hartert. XIV. Odonata', *Novit. zool.*, 20:468–469.

— (1921) 'The Odonata or Dragonflies of South Africa', *Annls South Afr. Mus.*, 18:245–452.

— (1924) 'Wissenschaftliche Ergebnisse der mit Unterstutzung der Akademie der Wissenschaften in Wien aus der Erbschaft Treitl unternommenen Zoologischen Expedition nach dem Anglo-Ägyptischen Sudan (Kordofan) 1914. Odonaten', *Denkschr. Akad. Wiss., Wien*, 99: 275–282.

— (1928) 'Zur Erforschung des Persischen Golfes. Libellen (Odonata)', *Wien. ent. Ztg*, 44:153–164.

Ris, F. & E. Schmidt (1936) 'Die *Pseudagrion* Arten des kontinentalen Afrika', *Abh. senckenb. naturforsch. Ges.*, 433:1–68.

Sage, B.L. (1960a) 'Notes on the Odonata of Iraq', *Entomologist*, 93:117–125.

— (1960b) 'Notes on the Odonata of Iraq', *Iraq nat. Hist. Mus. Publs*, 18:1–11.

Schmidt, Eb. (1978) Odonata, in: *Limnofauna Europaea*, 2nd Edition (J. Illies, ed.), G. Fischer, Stuttgart–New York, pp. 274–279.

Schmidt, E. (1915) 'Vergleichende Morphologie des 2 und 3 Abdominal-segments bei Männlichen Libellen', *Zool. Jb. Anat. Ontog.*, 39:87–200.

— (1929) Odonata, in: *Die Tierwelt Mitteleuropas*, Quelle u. Meyer, Leipzig, Bd. 4, Teil 1, lief 1b, 66 pp.

— (1938) 'Odonata aus Syrien und Palästina', *Sber. Akad. Wiss. Wien*, 1, 147:135–150.

— (1950a) 'Uber die Ausbildung von Steppenformen bei der Waldlibelle *Platycnemis pennipes* (Pall.)', *Ber. naturf. Ges. Augsburg*, 2:55–106.

287

— (1950b) 'Was ist *Libellula* .... *isoceles* O.F. Müller, 1767', *Ent. Z.*, 60:1–9.

— (1953) 'Zwei neue Libellen aus dem nahen Osten', *Mitt. münch. ent. Ges.*, 43:1–9.

— (1954a) 'Auf der Spur von Kellemisch', *Ent. Z.*, 64:49–62, 65–72, 74–86, 92–93.

— (1954b) 'Die Libellen Irans', *Sber. Akad. Wiss. Wien*, 1, 163:223–260.

— (1960) 'Betrachtungen an *Erythromma* Charp., 1840 (Odonata, Zygoptera)', *Gewässer & Abwässer*, 27:19–26.

— (1961) 'Ergebnisse der Deutschen Afghanistan-Expedition 1956 der Landessammlungen für Naturkunde Karlsruhe sowie der Expeditionen J. Klapperich, Bonn, 1952–1953 und Dr. K. Lindberg, Lund (Schweden) 1957–1960', *Beitr. naturk. Forsch. SüdwDtl.*, 19:399–435.

— (1968) 'Versuch einer Analyse der *Ischnura elegans* – Gruppe. (Odonata, Zygoptera)', *Ent. Tidskr.*, 88:188–216.

Schneider, W. (1981a) 'On a dragonfly collection from Syria', *Odonatologica*, 10:131–145.

— (1981b) 'Eine Massenwanderung von *Selysiothemis nigra* (Vander Linden, 1825) (Odonata: Macrodiplactidae) und *Lindenia tetraphylla* (Vander Linden, 1825) (Odonata: Gomphidae) in Südjordanien', *Ent. Z.*, 91:97–102.

— (1982a) 'Man-induced changes in the dragonfly fauna of the Jordan Valley', *Adv. Odonatol.*, 1:243–249.

— (1982b) '*Crocothemis sanguinolenta arabica* n. subsp. (Odonata: Anisoptera: Libellulidae), ein afrikanisches Relikt der südlichen Levante', *Ent. Z.*, 92:(3)25–31.

— (1983a) 'The larva of *Gomphus davidi* Selys, 1887', *Hydrobiologia*, 98:245–248.

— (1983b) 'Zur Eiablage von *Erythromma viridulum orientale* Schmidt 1960 (Odonata: Zygoptera: Coenagrionidae)', *Ent. Z.*, 93(16):255–299.

— (1983c) 'Désignation de lectotypes d'Odonates Levantines des collections du M.N.H.N., Paris, et description de la femelle de *Platycnemis kervillei* (Martin, 1909) [Odonata]', *Revue fr. Ent.*, (N.S.), 5(1):3–6.

— (1984a) 'Description of *Calopteryx waterstoni* spec. nov. from northeastern Turkey (Zygoptera: Calopterygidae)', *Odonatologica*, 13(2):281–286.

— (1984b) 'Beschreibung von *Gomphus kinzelbachi* n.sp. aus dem Iraq (Odonata: Anisoptera: Gomphidae)', *Ent. Z.*, 94(1/2):1–16.

— (1985a) 'Dragonfly records from SE-Turkey (Insecta: Odonata)', *Senckenberg. biol.*, 66(1/3):67–78.

— (1985b) 'Die Gattung *Crocothemis* Brauer 1868 im Nahen Osten (Insecta: Odonata: Libellulidae)', *Senckenberg. biol.*, 66(1/3):79–88.

— (1985c) 'Wiederbeschreibung von *Erythromma viridulum orientale* Schmidt 1960 aus dem östlichen Mittelmeerraum) (Insecta: Odonata: Coenagrionidae)', *Senckenberg. biol.*, 66(1/3):89–95.

— (1985d) 'The types of *Orthetrum anceps* (Schneider 1845) and the taxonomic status of *Orthetrum ramburii* (Selys 1848)', *Senckenberg. biol.*, 66(1/3):97–104.

— (1985e) '*Epallage fatime* (Charp.) (Zygoptera: Euphaeidae) as prey of *Argiope bruennichi* (Scop)) (Araneae: Araneidae)', *Notul. odonatol.*, 2(5):87.

— (1987a) 'The Genus *Pseudagrion* Selys, 1876 in the Middle East – A Zoogeographic Outline (Insecta: Odonata: Coenagrionidae)', in: *Proceedings of the Symposium on the Fauna and Zoogeography of the Middle East, Mainz 1985* (eds. F. Krupp, W. Schneider & R. Kinzelbach), TAVO A 28, pp. 114–123.

— (1987b) 'Die Verbreitung von *Onychogomphus macrodon* Selys, 1887, mit der Beschreibung des bisher unbekannten Weibchens und einer Wiederbeschreibung des Männchens (Odonata: Gomphidae)', *Opusc. zool. flumin.*, 13 (1987):1–12.

References

Schneider, W. & Z. Moubayed (1985) 'Beitrag zur Kenntnis der Odonata des Libanon', *Ent. Z.*, 95(13):183–192.

Schneider, W.G. (1845) 'Verzeichnis der von Herrn Prof. Dr. Loew im Sommer 1842 in der Türkei und Kleinasien gesammelten Neuropteren nebst kurzer Beschreibung der neuen Arten', *Stettin. ent. Ztg*, 6:110–116, 133–135.

Selys Longchamps, E. de (1837) *Catalogue des Lépidopterès ou Papillons de la Belgique, precédé du tableau des Libellulines de ce pays*. Desoer, Liège, 29 pp.

— (1839) 'Description de deux nouvelles espèces d'*Aeschna* du sous-genre *Anax* (Leach)', *Bull. Acad. r. Belg.*, S.1, 6:386–393.

— (1840) *Monographie des Libellulideés d'Europe*, Roret, Paris & Bruxelles, 220 pp.

— (1841) 'Nouvelles Libellulideés d'Europe', *Rev. zool. Soc. Cuvierienne*, 1841:243–246.

— (1848) 'Liste des Libellules d'Europe et diagnose de quatre espèces nouvelles', *Rev. zool. Soc. Cuvierienne*, 11:15–19.

— (1849) Libelluliens, in: Lucas, *Exploration scientifique d'Algerie, 3° partie, Animaux articulés. Neuroptera*, pp. 110–140, Paris.

— (1853) 'Synopsis des Calopterygines', *Bull. Acad. r. Belg.*, S.1, 20:1–73.

— (1862) 'Synopsis des Agrionines 2ᶜ Légion: Lestes', *Bull. Acad. r. Belg.*, S.2, 13:288–338.

— (1863) 'Synopsis des Agrionines 4ᶜ Légion: Platycnemis', *Bull. Acad. r. Belg.*, S.2, 16:147–176.

— (1867) 'Odonates recueillis à Madagascar et aux îles Mascareignes et Comores', in: *Recherches sur la faune de Madagascar* (eds. H. Schlegel & F. Pollen), Steenhoff, Leyden, 11 pp.

— (1871) 'Nouvelles revision des Odonates de l'Algérie', *Ann. Soc. ent. Belg.*, 14:9–20.

— (1876a) 'Synopsis des Agrionines 5ᵉ Légion: Agrion (suite), le grand genre *Agrion*', *Bull. Acad. r. Belg.*, S.2, 41:247–322; 496–539; 1233–1309.

— (1876b) 'Synopsis des Agrionines: (suite du genre *Agrion*) (Suite de la 5ᶜ Légion Agrion): Note additionelle et rectification au grand genre *Agrion*', *Bull. Acad. r. Belg.*, S.2, 42:490–531; 952–989; 989–991.

— (1878) 'Odonates de la région de la Nouvelle-Guinée', *Mitt. k. zool. Mus. Dresden*, 3:287–322.

— (1883) 'Synopsis des Aeschnines, 1ᶜ partie. Classification', *Bull. Acad. r. Belg.*, S.3, 5:712–748.

— (1884) 'Revision des *Diplax* palearctiques', *Annls Soc. ent. Belg.*, 28:29–45.

— (1887) 'Odonates de l'Asie mineure et Revision de ceux des autres parties de la faune dite européenne', *Annls Soc. ent. Belg.*, 31:1–85.

— (1891) 'Viaggio de Leonardo Fea in Birmania e regioni vicine 32. Odonates', *Ann Mus. civ. stor. nat. Genova*, S.2, 10:433–518.

Selys Longchamps, E. de & H.A. Hagen (1850) 'Revue des Odonates ou Libellules d'Europe', *Mém. Soc. r. Sci. Liège*, 6:xii + 408 pp.

Selys Longchamps, E. de & H.A. Hagen (1854) 'Monographie des Caloptérygines', *Mém. Soc. r. Sci. Liège*, 9:xi + 291 pp.

Selys Longchamps, E. de & H.A. Hagen (1857) 'Monographie des Gomphides', *Mém. Soc. r. Sci. Liège*, 11:viii + 460 pp.

Sowerby, F.W. (1917) 'Notes from Cairo', *Entomologist*, 50:9–11.

St. Quentin, D. (1964a) 'Die Odonaten der Sammelreise R. Petrovitz und F. Ressl aus Kleinasien', *Beitr. Ent.*, 14:421–426.

— (1964b) 'Ergebnisse der Von Dr. O. Paget und Dr. F. Kritscher auf Rhodos durchgeführten zoologischen Excursionen. XI. Odonata', *Annln naturh. Mus. Wien*, 67:659–660.

— (1964c) 'Odonaten aus Anatolien und dem Irak', *Ent. Mitt. zool. StInst. zool. Mus. Hamb.*, 3:49–51.

— (1965) 'Zur Odonatenfauna Anatoliens und angrenzenden Gebiete', *Annln naturh. Mus. Wien,* 68:531–552.

— (1968) 'Ergebnisse zoologischer Sammelreizen in der Türkei. Odonata', *Annln naturh. Mus. Wien,* 72:493–495.

Tennessen, K.J. (1975) 'Reproductive behaviour and isolation of two sympatric coenagrionid damselflies in Florida', *Diss. Abstr. B,* 36:5961.

Tillyard, R.J. (1917) *The Biology of Dragonflies,* University Press, Cambridge. 396 pp.

Tillyard, R.J. & F.C. Fraser (1938–1940) 'A reclassification of the order Odonata, based on some new interpretations of the venation of the dragonfly wing', *Aust. Zool.,* 9:125–169 (1938), 9:195–221 (1939), 9:359–396 (1940).

Valle, K.J. (1952) 'Die Odonatenfauna von Zypern', *Commentat. biol.,* 13:1–8.

Vander Linden, P.L. (1820a) 'Agriones bononienses descriptae', Bononiae, Tip. de Nobilibus, 8 pp.

— (1820b) 'Aeshnae bononienses descriptae, adjecta annotatione ad Agriones bononienses descriptas', Bononiae, Tip. de Nobilibus, 11 pp.

— (1823a) 'Agriones bononienses descriptae', *Opusc. Sci., Bologna,* 4:101–106.

— (1823b) 'Aeshnae bononienses descriptae, adjecta annotatione ad Agriones bononienses descriptas', *Opusc. Sci., Bologna,* 4:158–165.

— (1825) *Monographiae Libellulinarum Europearum Specimen,* Franck, Bruxelles, 42 pp.

Waage, J. (1979) 'Dual function of the damselfly penis: sperm removal and transfer', *Science,* 203:916–918.

Waterston, A.R. (1980) 'Insects of Saudi Arabia, Odonata', in: *Fauna of Saudi Arabia,* Vol. 2, pp. 57–70.

Watson, J.A.L. (1966) 'Genital structure as an isolating mechanism in Odonata', *Proc. R. ent. Soc. Lond.,* A, 41:171–174.

Williams, C.B. (1925) 'Notes on insect migration in Egypt and the near East', *Trans. R. ent. Soc. Lond.,* 1925:453.

— (1926) 'Further records of insect migration', *Trans. R. ent. Soc. Lond.,* 74:193–202.

INDEX

(ORDER ODONATA)

Valid names in roman type. Synonyms in italics. Page numbers in bold type indicate principal reference.

291

MAP

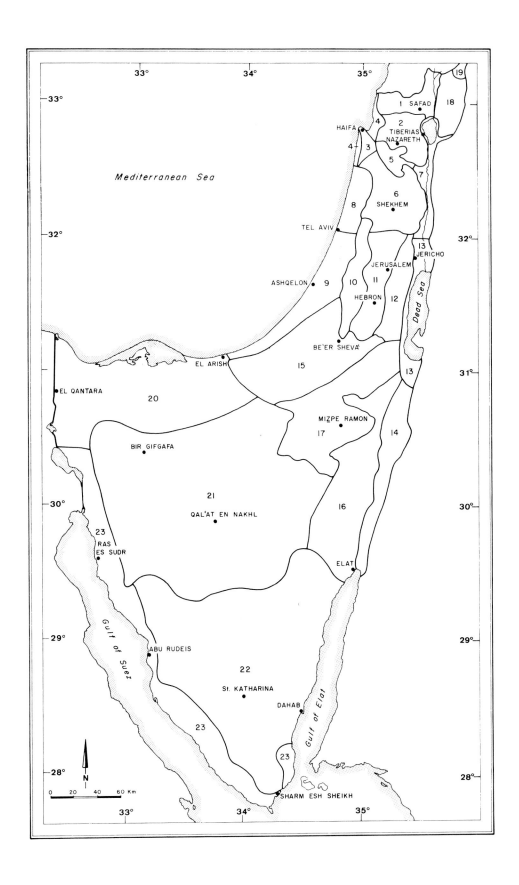

Geographical Areas in Israel and Sinai

KEY

1. Upper Galilee
2. Lower Galilee
3. Carmel Ridge
4. Northern Coastal Plain
5. Valley of Yizre'el
6. Samaria
7. Jordan Valley and Southern Golan
8. Central Coastal Plain
9. Southern Coastal Plain
10. Foothills of Judea
11. Judean Hills
12. Judean Desert
13. Dead Sea Area
14. 'Arava Valley
15. Northern Negev
16. Southern Negev
17. Central Negev
18. Golan Heights
19. Mount Hermon
20. Northern Sinai
21. Central Sinai Foothills
22. Sinai Mountains
23. Southwestern Sinai

כתבי האקדמיה הלאומית הישראלית למדעים

החטיבה למדעי-הטבע

החי של ארץ-ישראל

חרקים 5 — שפיראים
של קדמת המזרח התיכון
(ODONATA OF THE LEVANT)

מאת

הנרי דימונט

ירושלים תשנ"ב